Turfgrass Biotechnology: Cell and Molecular Genetic Approaches to Turfgrass Improvement

Edited by

Mariam B. Sticklen
Michael P. Kenna

Ann Arbor Press
Chelsea, Michigan

Editor's note: The galley proofs of each chapter have been confirmed by the senior author of each report. The scientific content of the book is the responsibility of the authors. However, the editors assume responsibility for informing the authors about any omissions and mistakes which may be found after the book is published. Thus, we kindly ask everyone to inform the editors about possible and necessary corrections.

Library of Congress Cataloging-in-Publication Data

Turfgrass biotechnology : cell and molecular genetic approaches to turfgrass improvement / edited by Mariam B. Sticklen, Michael P. Kenna.
p. cm.
Includes bibliographical references and index.
ISBN 1-57504-074-3
1. Turfgrasses--Biotechnology. I. Sticklen, Mariam B.
II. Kenna, Michael P.
SB433T825 1998
635.9′642233--dc21 97-22726
CIP

ISBN 1-57504-074-3

ANN ARBOR PRESS
121 South Main Street, Chelsea, Michigan 48118
Ann Arbor Press is an imprint of Sleeping Bear Press

Printed in the United States of America
1 2 3 4 5 6 7 8 9 0

Acknowledgments

The 1996 Turfgrass Biotechnology Workshop and this volume, *Turfgrass Biotechnology: Cellular and Molecular Genetic Approaches to Turfgrass Improvement,* have been supported by grants from the United States Golf Association, Michigan Turfgrass Foundation, Golf Course Superintendents Association of America, and Michigan State University Department of Crops and Soil Sciences and Michigan State Pesticide Research Center.

The editors owe an immense debt of gratitude to the authors who have contributed their time and efforts to this volume.

This book is dedicated to Mitra Sticklen and Susan Kenna for their patience, heartfelt support, and understanding.

Mariam B. Sticklen
Michael P. Kenna

About the Editors

Mariam B. Sticklen, Ph.D.

Dr. Mariam B. Sticklen is an Associate Professor in the Department of Crop and Soil Science and the Department of Entomology at Michigan State University. As the Principal Investigator of the MSU Agricultural Biotechnology for Sustainable Productivity Project, Mariam administered the ABSP Project as the Research Director from 1991 to 1994. Prior to her tenure at Michigan State University, she served as a faculty member at the Ohio State University and Clemson University.

Mariam has served at national and international levels on committees. At present, she serves as one of the members of the board of trustees of Consultative Group on International Agricultural Research (CGIAR) in Washington, DC and as one of the members of the board of governors of International Crops Research Institute for the Semi-Arid Tropics (ICRISAT). Mariam is also serving as a member of Faculty and Student Judiciary Committee and as a member of the Departmental Advisory Committee of the Crop and Soil Science Department at Michigan State University.

Mariam received her B.S. and M.S. degrees in horticulture, majoring in plant physiology. She received her Ph.D. in Horticulture, majoring in plant biotechnology from the Ohio State University in 1981. Mariam's research emphasis is on cellular and molecular genetic improvement of turfgrasses and other members of the grass family, including cereal crops. She has taught biotechnology courses, including "Methods in Genetic Engineering of Plants" and "Gene Expression." At present, Mariam teaches a graduate level course entitled "Ethical Responsibilities in Scientific Research" and an undergraduate course entitled "New Horizons in Biotechnology."

Mariam Sticklen has previously published a book entitled *Cell and Molecular Approaches to the Dutch Elm Disease Problem*. She has also published over sixty journal articles and book chapters. To date, Mariam has obtained 4 issued U.S. patents for her research in biotechnology.

Michael P. Kenna, Ph.D.

Dr. Michael P. Kenna has been the Director of USGA Green Section Research since February, 1990. His position with the United States Golf Association was created out of a need to extend greater administrative support to their growing turfgrass and environmental research program, which distributes over $1.5 million in grants annually. Mike travels extensively to visit turfgrass and environmental research sites and speak at conferences about the USGA's research programs, and serves on advisory boards and research foundations.

Mike received his B.S. degree in Ornamental Horticulture from California State Polytechnic University in Pomona, and while at Oklahoma State University received his M.S. degree in Agronomy and Ph.D. degree in Crop Science. His graduate studies involved turf and forage grass breeding, quantitative genetics, plant physiology, and turfgrass management. In 1985, Mike joined the faculty at Oklahoma State University as assistant professor, and was responsible for turfgrass research activities and a statewide extension program. He was selected for the Young Scientist position on the USGA Research Committee in 1988, and was invited to remain on the committee as a permanent member after his one-year term had expired. His academic background and familiarity with the workings of the Research Committee made him well suited for his current position with the USGA.

Mike is a native of San Diego, California, while his wife, Susan, is originally from Pasadena, California. Mike and Susan have been married for 17 years and have two sons, 11-year-old Patrick and 6-year-old Charles.

Introduction

The Apollo 11 mission to land on the moon is a memorable event for people around the world. My grandfather told me later it was hard for him to believe that the "Eagle" had landed and Neil Armstrong stepped out upon the moon. Why? My grandfather was born in 1900, before Orville and Wilbur Wright had successfully flown the *Flyer I* at Kitty Hawk, North Carolina, and now he had lived to see a person walk on the moon.

A person landing on the moon is truly a modern day highlight of what science and technology can achieve. The Smithsonian Museum has the *Flyer I* and Apollo 11 close enough to each other that, in an instant, one can see how far we have come. Unfortunately, other areas of science cannot display such a dramatic contrast so easily. I believe that recent developments in biotechnology are the biological equivalent of putting a person on the moon. However, we are just learning how to use this wonderful tool!

Over the last 75 years, a tremendous amount of effort has been exerted toward improving turfgrass species with conventional plant breeding techniques. During the last decade, we have just begun to scratch the surface of turfgrass improvement using cell and molecular techniques. These efforts have included:

- Research in genetic marker analysis and genetic mapping
- Biological control of turfgrass pathogens
- Manipulation and use of endophytes to improve turfgrass species
- *In vitro* culture of turfgrass species, including somaclonal variations and *in vitro* selections
- Cloning genes with potential to improve turfgrasses
- Studies of gene expression in turfgrasses
- Genetic engineering of turfgrass for abiotic and biotic stress resistance

Never before has an international group of scientists, working on different aspects of turfgrass improvement and biotechnology, had the opportunity to interact and collaborate with each other on the complex, multidisciplinary subject of biotechnology. This situation was the inspiration for this book, to bring together experts so they could share ideas, help solve potential problems, establish a network and identify research needs. In addition, golf associations, turf foundations, and industry leaders must recognize the need to support basic research that develops cell and molecular tools which aid the development and improvement of our important turfgrass species. As we go through this process, we need to keep asking where we want to be 5, 10, or 20 years from now!

The chapters in this book discuss four major areas where biotechnology promises to enhance breeding efforts developing turfgrasses that will meet the challenges of the next century. These areas include:

- Turfgrass and molecular marker analysis
- Biological control, including endophyte strategies
- Genes with potential for turfgrass improvement
- *In vitro* culture and genetic engineering

The rest of this introduction briefly highlights the information presented at this meeting.

Turfgrass Molecular Marker Analysis

Identifying the differences among varieties or cultivars of turfgrass species using molecular genetic analysis has made substantial progress in recent years. The techniques are faster and more accurate than techniques used just a short time ago. Molecular markers linked to important genes controlling a desirable trait should make breeding programs more effective. However, these molecular techniques have demonstrated that it is easier to prove things are different than to prove they are the same!

Genetic characterization of open pollinated turfgrass species will present a different set of problems compared to the characterization of asexually propagated cultivars. We need to know something about the breeding behavior and genetics of the turfgrass species to establish which genetic characterization techniques should be applied. Is the turfgrass species apomictic? Is it a homogeneous or heterogeneous population? Also, the statistical analysis techniques to help interpret the tremendous amount of data that can be produced from molecular analysis need further development. New techniques like AMOVA need to be implemented in order to interpret molecular analysis results.

Biological Control, Including Endophyte Strategies

The microorganisms that inhabit the turfgrass root zone are just starting to be characterized in a way that will lead to positive developments in biological control. It is important that the mechanism of biological control be understood thoroughly before we leap from the petri-dish to the field. The USGA has sponsored several projects in the area of biological control, and I believe that we are just starting to understand why some of these organisms fail when used in field situations.

The concept of *rhizosphere competence* is introduced and discussed in this section of the book. What is the biochemical mechanism by which the microorganism controls the pathogen? Is an antibiotic or siderophore produced by the organism? Does the organism parasitize the pathogen or does it induce a form of systemic resistance in the turfgrass plant?

How will these new biological controls be applied to turf? It appears that 10 billion (10^7) bacteria per gram of soil will be needed to achieve successful control of a turfgrass pathogen. Irrigation systems may be used to maintain high populations in turfgrass systems, but one must realize that this will require a closer relationship between the microbiologist, the plant scientist, and the agricultural engineer.

Fungal endophytes are already available in turf species such as ryegrass, tall fescue, and the fine fescues. An excellent summary describing the complexity of host plant and fungal interactions was prepared for this section of the book. One early approach to biological control was to inoculate endophyte-free turf species with a promising endophyte in the hope of establishing insect or disease resistance. Unfortunately, it will not be easy to move the existing endophytes from species to species. More basic research is needed in order to capitalize on this existing biological control method. And we must not forget, the choke or stromata formation problems which can cause health problems for both people and livestock need to be examined.

Genes with Potential for Turfgrass Improvement

This section of the book discusses research which has taken a molecular or biochemical examination of turfgrass response to an abiotic- or biotic-induced plant stress. Turfgrasses, like all plants, are *up* or *down* regulating genes in response to several plant stresses. Some of the examples include the production of chitinase during cold acclimation or the increased presence of desaturases which, through changes in the structure of the phospholipid fatty acid chains, improve the properties of the cell wall. In contrast to cold acclimation of warm-season species, the production of heat shock proteins in response to high temperatures may provide a means to improve adaptation of some cool-season species.

With regard to biotic stresses such as disease and insect problems, several research ideas are addressed in this section of the book. The research with the chitinase gene demonstrates that we can identify the DNA sequences of genes that may transfer or increase disease resistance. However, a great deal more needs to be understood about how these disease-resistant genes will work in an entirely different species. Also, since turfgrasses are perennial, will the introduction of a single gene produce long-lasting disease resistance? If not, how do we develop turfgrasses with several genes which confer resistance?

In Vitro Culture and Genetic Engineering of Turfgrasses

Tissue or *in vitro* culture already has proven to be a very useful tool in turfgrass breeding programs. In this last section, *in vitro* culture applications in conventional plant breeding programs is reviewed first. The articles following *in vitro* culture applications discuss the current genetic engineering research on important turfgrass species. In less than five years, these early efforts are very promising and have already produced turfgrass parental clones with commercial interest. In fact, efforts in the area of herbicide resistance have been a bit too successful for the companies who produce the herbicide products!

With regard to *in vitro* culture, the somaclonal variation present in many of the commercially important turfgrass species has been documented. The presence of this genetic variation has allowed turfgrass scientists and breeders an additional means to select for abiotic and biotic stress tolerance, as well as improve turfgrass quality.

The first chapter in this section presents how *in vitro* culture was used to produce interesting somaclonal variants of seashore paspalum The application of this technique on seashore paspalum is particularly useful because this open-pollinated species produces less than 5% viable seed and is self-incompatible. Conventional breeding methods are limited, and *in vitro* culture produced more than 4,000 regenerants that varied in genetic color, growth rate, density, and winter hardiness. Over 100 selections with improved turfgrass traits were selected for further evaluations.

Extensive selection and breeding of regenerated or transformed plants for agronomic characteristics is still required prior to their release as new cultivars. The second chapter in this section discusses the merits of using parental clones of existing cultivars as the explant material in order to provide the quickest and most rewarding approach. Work with the six parental clones of "Crenshaw" creeping bentgrass and five new zoysiagrass cultivars is presented.

Several forage and turfgrass species do not produce economically viable quantities of seed for commercial production. With the development of artificial seed technology through *in vitro* culture, it may be feasible to use this system to propagate parents for commercial production and release of hybrid varieties. In addition, if the parents of a commercial variety can be manipulated *in vitro,* the artificial seed method could facilitate the creation of varieties with value added traits introduced by genetic transformation. The third chapter in this section discusses these ideas and how they have been applied to embryo production in orchardgrass.

For turfgrass species, *in vitro* culture primarily has been used to produce somaclonal variants that are screened for an important characteristic under greenhouse or field evaluations. There is hope that *in vitro* culture can be used to actually select for superior somaclonal variants at the cellular level. *In vitro* selection for abiotic and biotic stresses is the focus of the next chapter in this section. This work has focused on selecting embryogenic, bentgrass callus for heat tolerance and disease resistance. The results of this effort have been very promising and developed HPIS, a novel *host plant interaction system* which allows the disease pathogen and turfgrass species to be cultured together *in vitro.*

The pioneering efforts in the new frontier of genetic engineering are discussed in the rest of this section. This research effort has been possible due to the advances made with important food and fiber crops. However, a pleasant surprise is how easily the success with agricultural crops was applied to turfgrass species. Several interesting examples of how transgenic turfgrass and forage clones with herbicide and disease resistance demonstrate the usefulness, and potential impact, that *biotechnology* will have on the turfgrass industry.

Aim for the Moon

In summary, I believe this book will someday be looked upon as the first serious, international and multidisciplinary discussion on the techniques and applications of biotechnology to greatly improve our turfgrass species. Independently, several scientists have made significant contributions in the areas of molecular marker analysis, biological control, identifying useful genes, *in vitro* culture and genetic engineering. The book provides the first rough map of the new turfgrass biotechnology frontier. My challenge to the reader is to work with others, conquer the remaining obstacles in our way, and aim for the moon!

Michael P. Kenna, Ph.D.

Contributing Authors

Michael P. Anderson
Department of Agronomy
369 Agriculture Hall
Oklahoma State University
Stillwater, OK 74078-0507

S. Assefa
Department of Agronomy
369 Agriculture Hall
Oklahoma State University
Stillwater, OK 74078-0507

Wm. Vance Baird
Horticulture Department
Clemson University
Clemson, SC 29634-0375

Faith C. Belanger
Center for Agriculture and Molecular Biology
Foran Hall, Dudley Road
Rutgers, Cook College
P.O. Box 231
New Brunswick, NJ 08903-0231

S.R. Bowley
Department of Crop Science
University of Guelph
Guelph, Ontario
Canada, N1G 2W1

Daniel C. Bowman
Department of Crop Sciences
2207 Williams Hall, Box 7620
North Carolina State University
Raleigh, NC 27695

Bruce E. Branham
Department of Natural Resources & Environmental Science
University of Illinois
W412 Turner Hall
Urbana, IL 61801

H. Brittain-Loucas
Crop Science Department
University of Guelph
Guelph, Ontario N1G 2W1
Canada

Stuart M. Brown
USDA, ARS
Plant Genetic Resources Conservation
Griffin, GA 30223-1797

Gustavo Caetano-Anollés
Ornamental Horticulture
The University of Tennessee
Knoxville, TN 37901-1071

Lloyd M. Callahan
Ornamental Horticulture and Landscape Design
The Institute of Agriculture
The University of Tennessee
Knoxville, TN 37901-1071

C.A. Cardona
University of Georgia
1109 Experiment Street
Griffin, GA 30223-1797

Benli Chai
Department of Crop & Soil Sciences
202 Pesticide Research Center
Michigan State University
East Lansing, MI 48824

T.K. Danneberger
Horticulture and Crop Science Dept.
202 Koffman Hall
2021 Coffey Rd.
The Ohio State University
Columbus, OH 43210-1086

Peter Day
Center for Agriculture and
Molecular Biology
Foran Hall, Dudley Road
Rutgers
Cook College
P.O. Box 231
New Brunswick, NJ 08903

A.R. Detweiler
Michigan State University
Pesticide Research Center
East Lansing, MI 48824-1311

J.A. DiMascio
202 Koffman Hall
The Ohio State University
Columbus, OH 43210-1086

D.S. Douches
286 PSSB
Department of Crops and
Soil Sciences
Michigan State University
East Lansing, MI 48824

Ron R. Duncan
College of Agric. & Env. Sciences
University of Georgia
1109 Experiment Street
Griffin, GA 30223-1797

N.M. Dykema
Michigan State University
Pesticide Research Center
East Lansing, MI 48824-1311

Milt C. Engelke
Texas Agriculture Experiment Station
Texas A&M University
17360 Coit Road
Dallas, TX 75252-9216

Shuizhang Fei
Department of Horticulture
377 Plant Science
University of Nebraska
Lincoln, NE 68583-0724

M. Gatschet
Department of Agronomy
369 Agriculture Hall
Oklahoma State University
Stillwater, OK 74078-0507

Farshid Ghassemi
Plant Molecular Genetics
The University of Tennessee
Knoxville, TN 37901-1071

Peter Gresshoff
Plant Molecular Genetics
The University of Tennessee
Knoxville, TN 37901-1071

R.K. Hajela
206 Pesticide Research Center
Department of Entomology
Michigan State University
East Lansing, MI 48824-1311

David R. Huff
Department of Agronomy
116 ASI Bldg.
Pennsylvania State University
University Park, PA 16802

Christopher A. Jester
USDA, ARS
Plant Genetic Resources
Conservation
Griffin, GA 30223-1797

Paul G. Johnson
Department of Horticulture
377 Plant Science
University of Nebraska
Lincoln, NE 68583-0724

J.C. Kamalay
USDA Forest Service
Northeast Exp. Station
Delaware, OH 43015

K.J. Kasha
Crop Science Department
University of Guelph
Guelph, Ontario N1G 2W1
Canada

Michael P. Kenna
USGA Green Section
P.O. Box 2227
Stillwater, OK 74076

Jeffrey V. Krans
Department of Plant & Soil Sciences
117 Dorman Hall
Mississippi State University
P.O. Box 9555
Starkville, MS 39759

Stephen Kresovich
USDA, ARS
Plant Genetic Resources Conservation
Griffin, GA 30223-1797

Lisa Lee
The Scotts Company
14111 Scottslawn Rd.
Marysville, OH 43041

Chien-An Liu
Department of Crops and
Soil Sciences
202 Pesticide Research Center
Michigan State University
East Lansing, MI 48824

Zhao Wei Liu
USDA, ARS
Plant Genetic Resources
Conservation
Griffin, GA 30223-1797

D.S. Luthe
Dept. of Biochemistry and
Molecular Biology
Box 9650
Mississippi State University
Mississippi State, MS 39762

B.D. McKersie
Crop Science Department
University of Guelph
Guelph, Ontario N1G 2W1
Canada

S.E. Mitchell
USDA, ARS
Plant Genetic Resources
Conservation
Griffin, GA 30223-1797

M.G. Nair
Michigan State University
Pesticide Research Center
East Lansing, MI 48824-1311

Eric B. Nelson
Department of Plant Pathology
334 Plant Sciences Bldg.
Cornell University
Ithaca, NY 14853

S.L. Park
Department of Biochemistry and Molecular Biology
Box 9650
Mississippi State University
Mississippi State, MS 39762

Donald Penner
Department of Crops and Soil Sciences
470 PSSB
Michigan State University
East Lansing, MI 48824

Ingo Potrykus
ETH
Institut für Pflanzenwissenschaften
ETH-Zentrum
CH-8092 Zürich
Switzerland

Gary L. Powell
Department of Biological Sciences
Clemson University
Clemson, SC 29634

J.F. Powell
Michigan State University
Pesticide Research Center
East Lansing, MI 48824-1311

Paul E. Read
Department of Horticulture
377 Plant Science
University of Nebraska
Lincoln, NE 68583-0724

B. de los Reyes
Department of Agronomy
369 Agriculture Hall
Oklahoma State University
Stillwater, OK 74078-0507

Michael D. Richardson
Department of Plant Science
Foran Hall
Rutgers University
Cook College
New Brunswick, NJ 08903

Melissa B. Riley
Department of Plant Pathology and Physiology
Clemson University
Clemson, SC 29634

Terrance P. Riordan
Department of Horticulture
377 Plant Science
University of Nebraska
Lincoln, NE 68583-0724

Suresh Samala
Horticulture Department
Clemson University
Clemson, SC 29634-0375

Germán Spangenberg
Plant Sciences and Biotechnology
Agriculture Victoria
La Trobe University
Bundoora, Victoria 3083
Australia

Mariam B. Sticklen
Department of Crop & Soil Sciences
206 Pesticide Research Center
Michigan State University
East Lansing, MI 48824-1311

Patricia M. Sweeney
Horticulture and Crop Science Dept.
202 Koffman Hall
2021 Coffey Rd.
The Ohio State University
Columbus, OH 43210-1086

C. Taliaferro
Department of Agronomy
369 Agriculture Hall
Oklahoma State University
Stillwater, OK 74078-0507

B. Tar'an
Crop Science Department
University of Guelph
Guelph, Ontario N1G 2W1
Canada

M. Tomaso-Peterson
Department of Plant & Soil Sciences
Box 9555
Mississippi State University
Mississippi State, MS 39762

Joseph M. Vargas, Jr.
Dept. of Botany & Plant Pathology
Michigan State University
Pesticide Research Center
East Lansing, MI 48824-1311

Zeng-yu Wang
Plant Sciences & Biotechnology
Agriculture Victoria
La Trobe University
Bundoora, Victoria 3083
Australia

Donald Warkentin
206 Pesticide Research Center
Department of Entomology
Michigan State University
East Lansing, MI 48824-1311

Scott E. Warnke
National Forage and Turf Seed
Production Research Center
3450 S.W. Campus Way
Oregon State University
Corvallis, OR 97333

John Wells
Horticulture Department
Clemson University
Clemson, SC 29634-0375

James F. White, Jr.
Department of Plant Pathology
Foran Hall
Rutgers University
Cook College
New Brunswick, NJ 08903

Ikuko Yamamoto
Texas A&M Research and
Extension Center
17360 Coit Road
Dallas, TX 75252

Jiyu Yan
Valdosta State University
Valdosta, GA 31698

Heng Zhong
Department of Crop and
Soil Sciences
286 PSSB
Michigan State University
East Lansing, MI 48824

Contents

Part 1
Turfgrass Molecular Marker Analysis

Part 2.
Biological Control, Including Endophyte Strategies

Part 3.
Genes with Potential for Turfgrass Improvement

Part 4.
In Vitro Culture and Genetic Engineering of Turfgrass

Part 5.
Future Perspectives

Part 1

Turfgrass Molecular Marker Analysis

Chapter 1

Molecular Genetic Analysis of Turfgrass

P.M. Gresshoff, L.M. Callahan, F. Ghassemi, and G. Caetano-Anollés

INTRODUCTION

Science in general deals with the discovery of natural laws which permit the prediction of some events, and thus allow proactive decision making. This motive also exists in genetic science, which seeks to understand the laws of inheritance, the mechanisms of gene expression, and the complexities of environment-genotype interactions. Genetics has been exceedingly successful in producing productive animal and plant genotypes, which are used in agriculture, silviculture, and horticulture. Genetics has provided tools to diagnose and explain genetic disorders. This has led to the field of prenatal diagnosis and genetic counseling in human medicine (Rommens et al., 1989). Because of an understanding of the molecular causes of a genetic condition, steps can be taken to lessen the impact. For example, in medicine, direct drug design takes advantage of three-dimensional modeling of target receptors or enzymes, for which topographically correct ligand molecules can then be designed.

The understanding brought about by molecular biology and genetics has opened the field of biotechnology. Whilst the exploitation of biological processes to yield quality products is as old as humanity (c.f., beer, cheese, and wine making), the advent of recombinant DNA technology allowed the quantum leap, as gene sequences previously harbored in different organisms could be combined. Biotechnology has many different definitions; for the purpose of this chapter I have chosen a broad one, as shown in Figure 1.1.

Biotechnology today comprises perhaps four main areas. These are: gene transfer (also commonly called genetic engineering using recombinant DNA technology), molecular diagnostics (sometimes called fingerprinting, genetic

Biotechnology is:
the directed application of
technologies to select or develop
novel heritable genetic properties
in organisms or
their cellular processes to
generate products for the
benefit of humankind and the planet

Figure 1.1. A broad definition of biotechnology. More restrictive terms could include the terms genetic engineering, recombinant DNA technology, and "not normally found in Nature."

profiling, DNA forensics), cell biology (as used in embryo rescue from wide crosses, tetraploidization to increase vigor, phytohormones, and growth regulators) and informatics (databases, three-dimensional molecular modeling, communication networks, image and pattern analysis, and robotics). See Figure 1.2.

Plant Biotechnology as Related to Turfgrasses

To assist the genetic improvement of turfgrasses, two basic technologies, namely gene transfer and gene diagnostics, are needed.

Gene transfer technologies have reached a high level of sophistication and economic success. It is worthwhile to remember that the first recombinant DNA molecules were only produced less than 25 years ago. The first verified gene transfer in plants was achieved about 10 years ago, when scientists in several companies — such as Monsanto, Agrigenetics, and Ciba Geigy — achieved the simple transfer of a kanamycin resistance gene to tobacco or sunflower.

Since then major successes have been achieved. In this summer of 1997, U.S. agriculture has access to transgenic corn, having been made resistant to corn borer by the transfer of a bacterial toxin protein gene encoding the BT endotoxin. Similar material is available in soybean and cotton. Virus-resistant squash plants and potato are available after the transfer of coat protein genes, which blocks viral replication. For the last three years we have had governmental approval for a value-added crop, namely the Flavr Savr® tomato, which contains an antisense copy of a cell wall–softening enzyme, permitting longer

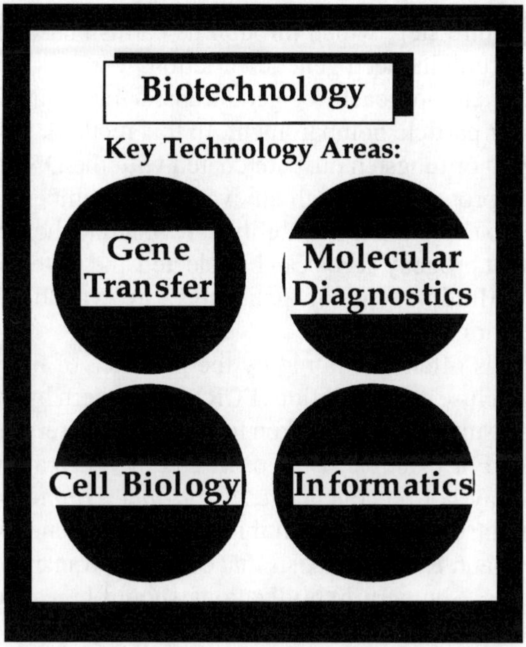

Figure 1.2. Key component areas of biotechnology as discussed in this chapter. Informatics deals with the immense computer and information-based revolution occurring presently, which allows the storage and comparison of large data sets. Other than a peripheral mention of map-making programs, this area will not be discussed, as it is beyond the expertise of the author.

growth and flavor development on the vine. Several crop plants have been genetically engineered for herbicide resistance. Most prominent is the Round-Up resistance in soybean, engineered through Monsanto Agricultural Chemical Company (or CERAGEN Inc., as it is called now for the biotechnologically engineered plant products). Various other examples exist, dealing with disease resistance, value-added features, and horticultural properties. For example, a branch of Calgene Inc. working in Australia has recently used genetic engineering to produce a blue flowering rose. Also, Round-Up resistance was transferred to poplar trees.

Gene transfer is based on two fundamental technologies. The first uses the natural transfer of DNA from the bacterium *Agrobacterium tumefaciens* or *Agrobacterium rhizogenes* to the plant. New DNA sequences can be fused to the naturally transferred T-DNA (transfer DNA) to facilitate the gene transfer. Sophistication of this approach involves the use of binary vector, selectable markers, and reporter genes to monitor gene transfer, and the use of tissue-specific or inducible promoters, as well as varying methods of co-culture, vacuum infiltration, or simple stabbing. Often the *Agrobacterium* route involves a tissue culture step, which through its callus phase may increase the amount of cell culture–induced genetic variation.

The second broad approach for gene transfer into plants uses a process called biolistics, or particle bombardment. In this method, small (1 μm diameter) pieces of gold or tungsten dust are coated with the DNA of interest. The particles are then propelled through an evacuated chamber into plant tissue. This shotgun approach requires the ability to regenerate the target tissue into a fertile plant, so that progeny seeds can be selected and tested for the effects of the gene transfer. Multiple types of "gene guns" are available, as are protocols for plant regeneration.

Gene transfer is often monitored by the presence of the gene of interest through the polymerase chain reaction (PCR) or Southern hybridization. Some precautionary measures should be taken to avoid false interpretations of transformation results. For example, PCR or Southern verification using internal gene fragments may not be conclusive in *Agrobacterium*-based transformations, as the bacteria, despite strong antibiotic selection, tend to maintain themselves inside the plant. Evidence exists that even seed transmission is possible. To circumvent this, Southern hybridizations should be carried out on undigested DNA to monitor for high-molecular-weight DNA integrations larger than the plasmid size of the transfer construct. At times monitoring for border fragments is advisable. The use of a reporter gene interrupted by a plant intron is an elegant way to verity plant transfer, as the bacterium cannot excise the intron and express the gene.

DNA Diagnostics and Molecular Markers

The ability to clone a piece of DNA has opened the possibility to use such clones as probes to study the genetic makeup of an organism. In part this resulted in the explosion of gene sequence information. For example, for the model plant *Arabidopsis thaliana* (a crucifer) about 20,000 cDNA sequences have been characterized; for rice (*Oryza sativa*), owing to a large effort by the Japanese research consortium at Tsukuba, Japan, nearly 30,000 gene sequences are known.

When such cloned gene fragments are used as probes to detect restriction fragment length polymorphisms (RFLPs), they are referred to as ESTs (expressed sequence tags). ESTs have the advantage that in some circumstances, phenotypic traits are co-mapped to the same region. In that case, the EST clone is a candidate to be causative for the phenotype.

Alternatively, random genomic clones are used as probes (see Landau-Ellis et al., 1991). Usually these are from the same organism, but often interspecific homology exists, allowing "cloning by phone."

Recently the spectrum of molecular markers expanded because of the application of PCR and arbitrary primer technology. Currently used molecular markers are RFLPs (either ESTs or genomic clones), PCR markers, arbitrary primed PCR markers (DAF, RAPD, AP-PCR, AFLP), microsatellites and isozymes. The latter involves analysis of proteins and is relatively limited, as the degree of polymorphism and the number of loci is small.

Arbitrary primed PCR approaches are more efficient than RFLPs in detecting molecular polymorphisms; they have a higher multiplexing ratio. These approaches are based on the PCR, but involve significant modifications. The essential fact is the nature of the primer. In arbitrary primer technology, the primer is a short oligonucleotide, usually, but not necessarily, used in singularity, but of an arbitrary sequence (Caetano-Anollés et al., 1991; Bassam and Bentley, 1995). This targets multiple regions of the genome concurrently. Three methods were developed more or less concurrently, and are covered by separate U.S. patents. The most commonly used is the RAPD method developed by DuPont Company (Williams et al., 1990). Similar is the AP-PCR method discovered by Welsh and McClelland (1990). The DAF (DNA amplification fingerprinting; U.S. patent No. 5,413,909) method was developed with the shortest primers and a larger number of amplification products (Caetano-Anollés et al., 1991; Caetano-Anollés and Gresshoff, 1994c). DAF can be used at higher annealing temperatures (Caetano-Anollés et al., 1992), uses a high primer concentration (3 μ*M* for linear primers and up to 30 μ*M* for structured mini-hairpin primers (Caetano-Anollés and Gresshoff, 1994a, 1996), and low (50 pg/μL) template DNA levels. DAF profiles are usually generated by separation of the amplification products on thin (0.45 mm) polyacrylamide (10%) gels backed by Gelbond plastic film support, which maintains the gel integrity and permits efficient storage of air-dried gels. DAF gels are commonly stained with a fast and efficient silver staining procedure (Bassam et al., 1991; Caetano-Anollés and Gresshoff, 1994b). The silver staining procedure is covered by U.S. patent No. 5,643,479, and is commercially available through the SilverSequencing kit marketed by Promega, Inc.

Weaver et al. (1994) described the reisolation of the DNA fragment from such polyacrylamide gels using reamplification and cloning. Such cloned bands were used for DNA sequence determination (Prabhu, 1995) as well as probes for DAF gel blots or genomic Southern blots. Cloned bands contain the primer sequence at both terminal regions, and most frequently detect unique DNA sequences in genomic Southern hybridizations. The similarities and differences between the arbitrary primer methods have been reviewed (Gresshoff, 1994; Caetano-Anollés and Gresshoff, 1994a,c).

While amplification methods are susceptible to variation stemming from thermocycler parameters, DNA concentration, and quality, Gresshoff and Mackenzie (1994) detected only small variation in amplification profiles in a study of soybean. It is presumed that this robustness stems from the optimized DAF conditions pertaining to high-primer:low-template DNA concentration, the use of Stoffel DNA polymerase, and the application of PAGE and silver staining, which permit detection of aberrant amplifications (Gresshoff, 1994; Caetano-Anollés and Gresshoff, 1994a,c). Recently, we used 55°C annealing with 8-mer, 10-mer and mini-hairpin primers to achieve robust and high-resolution DAF.

DAF analysis allowed the distinction of plant genotypes for pedigree analysis (Prabhu et al., 1997) and phylogenetic comparisons (Caetano-Anollés et al., 1995; Weaver et al., 1995). For example, 10 soybean lines were supplied for a determination of their relatedness. Many of them were breeding lines of known pedigree relation. Seven DAF primers distinguished nine genotypes; with prior knowledge, only three primers were required. The resultant dendrogram of relatedness matched that proposed by the breeders, as well as that produced by RFLP analysis. However, the RFLP dendrogram required 53 selected probes to obtain the data set. Clearly, arbitrary primer technology is more efficient, because of its high multiplex ratio.

Recently another multiplex technology was developed by KeyGene Inc. in the Netherlands (Vos et al., 1995). The technology, called selective restriction fragment amplification, or commercialized under the acronym AFLP,[1] generates characteristic DNA profiles from restriction fragments, which were ligated to two distinct adapters. PCR primers, specific to the adapter sequences, but extended at their 3′ end by 3 nucleotides, amplify a subset of restriction fragments. About 100–120 products are usually separated by PAGE and visualized by automated DNA sequencing or autoradiography. Optimally, separation works with 70–80 bands, numbers which can also be achieved with DAF using mini-hairpin primers and long gels. Markers produced by AFLP are predominantly (85–90%) dominant; rehybridization to genomic blots suggests that most represent repeated DNA (10–20 copies). Some suggestions exist that AFLP markers are predominantly clustered in areas of genetic maps representing perhaps centromeric heterochromatin, which would be consistent with the highly repeated nature of the marker. Whether such trends are caused by primer design favoring repeated DNA or are a reflection of the restriction nucleases used to generate the original fragments remains to be tested. Several commercial kits are now available for AFLP analysis, some of which couple the technology to semiautomated gel separation in DNA sequencers.

Other forms of molecular markers exist. Of these, single-sequence repeats (also called SSRs or microsatellites) are of great value. Akkaya et al. (1992) and Weising et al. (1989, 1992) demonstrated their application to plant breeding. Because they are based on a PCR strategy, only a small amount of starting material is needed. Furthermore, they tend to exist as loci, giving a greater resolution power, and are codominant (Rongwen et al., 1995). However, they have one major limitation, because SSRs require substantial prior knowledge. This is made easily available for elite species such as soybean, humans, *Arabidopsis,* and maize. However, fringe species will not be accessible in the near future because of the lack of sufficient numbers of SSR loci. Additionally, there may be a population genetic question relating to the utility of SSRs.

---

[1] The term AFLP was originally used by Caetano-Anollés et al. (1991) to describe polymorphisms generated by DAF. However, because of broad usage of the term in relation to SRFA, and in an attempt to avoid confusion, its connection to DAF is discouraged.

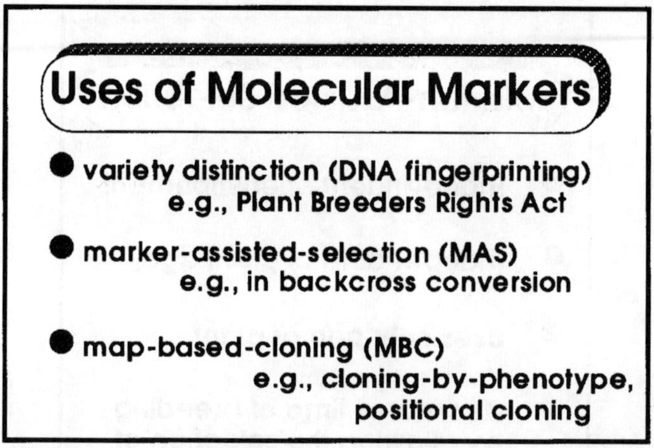

Uses of Molecular Markers
variety distinction (DNA fingerprinting)
e.g., Plant Breeders Rights Act
marker-assisted-selection (MAS)
e.g., in backcross conversion
map-based-cloning (MBC)
e.g., cloning-by-phenotype,
positional cloning

Figure 1.3. Some uses of molecular markers. Markers may be RFLPs (see Landau-Ellis et al., 1991), PCR markers, arbitrary primed markers (Caetano-Anollés et al., 1991a; Williams et al., 1990; Welsh and McClelland, 1990), SSRs (Akkaya et al., 1992), AFLPs (Vos et al., 1995), or isozymes.

They tend to be found in expressed sequences, and have been shown to be the cause for some genetic diseases in humans (such as the propensity to develop bowel cancer). Thus the polymorphism may not be "neutral" and alleles may exist at different frequencies in different populations.

Molecular markers have diverse uses (Figure 1.3). They are commonly used for variety distinction and plant variety rights issues. Likewise the genotyping allows quality control in biological material, which is especially useful when phenotypically related material is grown, shipped, and marketed concurrently.

Molecular markers serve in marker-assisted selection (MAS; Webb et al., 1996). This permits the detection of a chromosomal region by molecular means. If this region harbors a gene controlling a morphological, disease resistance, or agronomic phenotype, detection of the marker alleviates the need for phenotypic testing (Figure 1.4). This requires extensive mapping in the first case, and demonstrates why molecular marker systems with high data content (such as SSRs, DAF, and AFLPs) are more convenient than RFLPs. MAS is practical for backcross conversion, as multiple regions of the genome of a donor plant can be introgressed into a new variety (Gresshoff, 1995b).

The third utilization of molecular markers is in gene discovery through map-based cloning or positional cloning or cloning by phenotype (Young and Phillips, 1994; Wicking and Williamson, 1991). This approach takes advantage of close linkage between a molecular marker and a mapped gene (Gresshoff, 1995a). The marker is used as a probe to isolate large DNA molecules cloned into new cloning vectors (pYAC4 and pBACBeLo11) to make

Marker-Assisted Selection

- independent of environment
- independent of plant age
- uses only part of plant
- decreases time of breeding by allowing the selection of preferred backcross plants

Figure 1.4. Marker-assisted selection increases the power of classical plant breeding by defining regions of parental genomes.

either yeast artificial chromosomes (YACs) or bacterial artificial chromosomes (BACs), respectively (Zhu et al., 1996; Pillai et al., 1996; Woo et al., 1995; Shizuya et al., 1992). These large pieces of DNA (as large as 900 kb fragments have been isolated in our laboratory) permit the search for mutant versus wild-type gene sequences associated with the phenotype of interest (Gresshoff, 1995). Spectacular success was achieved using this approach for the isolation and characterization of plant disease-resistance genes (Mindrinos et al., 1994; Bent et al., 1994; Martin et al., 1993; Song et al., 1995).

Isolated genes can serve as useful markers for turfgrass molecular biology because of sequence conservation. Likewise, molecular maps generated for other monocotyledonous plants such as rice and maize can serve as guides for marker arrangement in turfgrasses, because of a phenomenon called synteny (Moore et al., 1993). It is now clear that many plant groups share broad genomic blocks, so that the arrangement of molecular markers is maintained. This was best demonstrated in the cereals, where rice, corn, wheat, barley, sorghum, and rye show strong synteny. Similar features are found in legumes (soybean, cowpea, Frenchbean, and mungbean) and crucifers (*Arabidopsis* and rape seed).

Synteny and EST homology coupled with high-resolution marker technology provide hope for plant species that are outside the mainstream of research activity. Information and material can be transferred in an ever increasing information environment. The reason for isolating genes is to be able to predict their function. Isolated genes also come with their regulatory regions, called promoters, which function as tools for directed expression in gene transfer experiments. For example, a promoter that is only functional in the root may be a good candidate to transfer a toxin gene effective against root nematodes. Isolated genes can be modified *in vitro* and reintroduced into plants to provide new physiological functions (c.f., Flavr Savr® tomato).

Molecular Markers Used to Identify Problems with Turfgrasses

In turfgrass we have a knowledge gap compared to the agronomically important cereals; hence we are in a "catch-up" mode. As stated above, EST homology and synteny may be of great benefit. To initiate our research with turfgrasses, we focused on DNA marker technology for variety distinction, quality control, and to lay the basis for marker-assisted selection (MAS). To achieve the latter, molecular diagnostics need to be coupled to an existing turfgrass breeding program to detect linkage of traits and markers. Through MAS, plant breeding becomes proactive, as easily detectable markers are closely associated with horticulturally important characteristics affecting appearance, performance, disease, and stress responses.

We used the high-resolution, nonradioactive DNA amplification fingerprinting (DAF) method to analyze the genomes of centipedegrass and bermudagrass. Other grasses, such as creeping bentgrass, *Zoysia*, bluegrass, and St. Augustinegrass were also analyzed. For turfgrass analysis (see Weaver et al., 1994, 1995; Caetano-Anollés et al., 1995, 1997), DAF usually uses short arbitrary primers of 8 nucleotide length, or structured mini-hairpin primers (Caetano-Anollés et al., 1996), in which a stem-loop structure is attached on the 5′ end. The minimum 3′ unstructured extension is three nucleotides. Such primers detect high degrees of variation. In a typical experiment, template DNA (about 50–100 pg/μL) is amplified with AmpliTaq Stoffel fragment DNA polymerase, 3 μM primer, using a 2-step temperature 35-cycle program. Amplification products are separated by polyacrylamide gel electrophoresis in 0.45-mm-thick gels, then stained with silver to produce permanent DNA profiles. The potential to detect variation is nearly limitless. The procedure offers the advantage of fast data analysis, permanence of the gels and amplification products, absence of radioactivity, high resolution, and high information content per analysis. Typically, about 40 products are resolved per primer. Semiautomation of analysis is possible through the use of a PhastSystem or capillary electrophoresis.

We have used DAF to study the genetic variation of bermudagrass (*Cynodon*) species and cultivars of interspecific crosses that exhibit leaf blade textural characteristics ranging from coarse to fine (Caetano-Anollés et al., 1995). Arbitrary octamer primers produced complex and reproducible amplification profiles with high levels of polymorphic DNA. Phylogenetic analysis using parsimony and unweighted pair (PAUP) group cluster analysis using arithmetic means (UPGMA) grouped 13 bermudagrass cultivars into several clusters, including one containing the African bermudagrasses (*C. transvaalensis*) and another containing the common bermudagrasses (*C. dactylon*). The latter group included *C. magennissii* ("Sunturf") and an interspecific *C. transvaalensis* X *C. dactylon* cross ("Midiron") — two cultivars that exhibited leaf textural characteristics closer to the common types. All other *C. transvaalensis* X *C. dactylon* crosses grouped between the African and common types. An extended screen of 81 octamer primers was needed to separate cultivar "Tifway" from the irradiation-induced mutant "Tifway II." The use of template endonuclease digestion prior to amplification or arbitrary mini-hairpin primers increased detection of polymorphic DNA and simplified the task of distinguishing these closely related cultivars. Alternatively, the use of capillary electrophoresis (CE) resolved fingerprints adequately and detected products with high sensitivity, promising to increase throughput and detection of polymorphic DNA. When used to fingerprint samples from commercial sources, DAF identified bermudagrass plant material based on unique reference profiles generated with selected primers.

Bermudagrasses are frequently complicated by what is known as "off-types." There are several explanations for these, the major ones being contamination with other genotypes and spontaneous mutations. The mechanism for such mutational change has not been elucidated, but may involve retrotransposons, deletions, satellite DNA expansion, aneuploid, translocation, and point mutations. It is possible that the triploid condition, involving an interspecific hybrid, increases the frequency of such events. At the same time, contamination from other grass lines is possible during vegetative production at sod farms. It seems essential that more research be conducted into this problem. Of interest is the fact that Tifway 419, when DAF fingerprinted from several university sources, gave identical results. In other words, Tifway 419 does not seem to suffer from spontaneous variation if kept under small-scale propagation. However, despite this fact, our laboratory repeatedly receives material from growers and users, who have off-types in Tifway material (see Figures 1.5a and b).

Figures 1.5a and b show an example of DAF patterns generated from Tifway 419 and Tifway II. Primers OR30 and OR33 were used to determine whether seven turfgrass samples from a golf course in Florida were indeed Tifway. Prior DAF analysis using Tifgreen 328 and Tifdwarf indicated that Samples 1a to 7a were not the same as the reference checks, and that Sample 5a was more distinct, while the remaining samples were apparently identical to each other. The present analysis showed that indeed Sample 5a is really distinct, while the other samples are similar if not identical to each other. Most importantly, none of them are related to Tifway.

In another study, DAF was used to characterize bermudagrass (*Cynodon*) species and cultivars of interspecific crosses, evaluate genetic diversity and origin of bermudagrass off-types, and certify authenticity of cultivar stocks (Caetano-Anollés et al., 1997). An initial analysis of 13 bermudagrass cultivars was used as a reference to establish genetic relationships between culti-

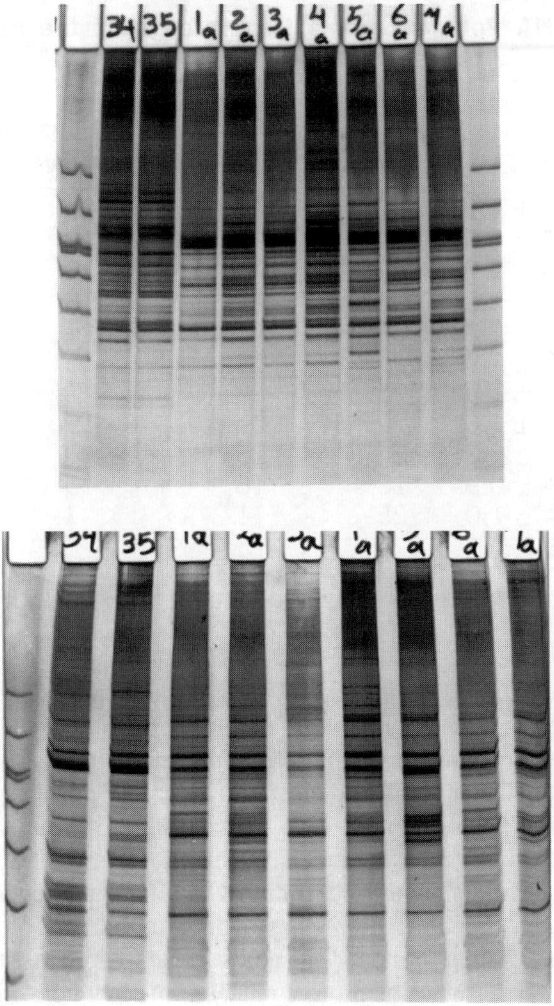

Figure 1.5. DAF patterns separated on thin polyacrylamide gels, stained with silver. Panel (a) separation using primer OR30. Panel (b) using primer OR33. Molecular size markers range from 100 bp to 1000 bp. #34 = Tifway 419; #35 = Tifway II. Both obtained from Athens, Georgia. 1a to 7a are bermudagrass samples collected by P.M.G. at a golf course in Florida.

var Tifway 419 and off-types that produce patches exhibiting contrasting morphology and performance. Phenetic analysis using cluster (UPGMA) and ordination (PCO) techniques showed that Tifway 419 was genetically stable, and that off-types were genetically distant to Tifway 419, representing a heterogeneous group of bermudagrass probably of interspecific hybrid origin. While Tifway 419 standards from different sources could not be differentiated

Table 1.1. DNA Polymorphisms Generated from thc Analysis of 43 Bermudagrass Cultivars, Accessions, and Off-Types.

Primer	Avg. No. of Bands/Primer (± SE)	DAF Loci	Polymorphic Loci	Loci within Tifway[a]	% Polymorphic Loci
8-2	19.6 ± 2.1	29	23	22	79
8-4	15.3 ± 1.8	24	19	18	79
8-5	14.0 ± 1.7	24	24	20	100
8-8	12.4 ± 1.3	25	22	18	88
8-9	15.4 ± 2.2	31	28	25	90
8-10	13.8 ± 1.5	25	19	15	76
8-12	15.7 ± 1.9	25	22	18	88
8-16	12.9 ± 1.6	22	21	19	96
8-20	19.2 ± 1.7	24	15	10	60
8-23	11.4 ± 0.9	18	16	10	89
8-26	22.4 ± 1.0	25	11	7	44

[a] Number of amplification products of less than 500 bp that are polymorphic within reference, standard, and off-type Tifway plant material.

from each other, the off-types were diverse at the genetic level, their origin being clearly from sod contamination and not from somatic mutation. DAF was used to fingerprint samples in "phytoforensic" applications, identifying bermudagrass plant material based on unique reference profiles generated with selected primers. Exclusions were observed in 63% of 93 bermudagrass samples believed related to cultivars Tifway, Tifgreen and Tifdwarf. Conversely, calculation of fingerprint matching coefficients facilitated inferences on the probability of a match in inclusion cases as applied to cultivar verification, seed certification, and cultivar protection (Tables 1.1 and 1.2). This study presented the foundation for the identification of mistakes in plantings, mislabeled plant materials, and contamination or substitutions of sod fields.

Molecular genetic analysis of turfgrasses has just begun. Many obstacles exist, although the plant group is so significant for recreation, ecology, and beautification. The genetics of several grasses are difficult, and uncertainties exist about genetic stability. Moreover, federal funding sources view turfgrass as a tangential plant group, and believe that commodity and interest groups, such as superintendents and the USGA, should focus on research support. Because of the absence of significant research commitment over the last 10 years into molecular biology and genetics of turfgrasses, the field is in a catch-up mode. Several key experiments need to be done, and milestones need to be achieved. Through our initial work using DAF analysis we have laid some of the foundation for further work. We have shown that technology that was designed for other crops can easily be applied to turf. We hope that the industry sees the value of this technology and supports active interaction among university, governmental, and private laboratories.

Table 1.2. DNA Polymorphisms Generated from the Analysis of 13 Bermudagrass Cultivars.

Primer	Oligonucleotide Sequence (5' to 3')	DAF Loci[a]	Polymorphic Loci	No. of Unique Fingerprints[b]	Band Sharing Frequency (x)[c]	Probability of a Match[d]
8·2	CCTGTGAG	37	32	12	0.567	$6.8 \cdot 10^{-6}$
8·4	GTAACGCC	26	20	11	0.530	$1.6 \cdot 10^{-4}$
8·5	GACGTAGG	32	27	11	0.563	$3.2 \cdot 10^{-5}$
8·8	GAAACGCC	29	23	10	0.576	$1.0 \cdot 10^{-4}$
8·9	GTTACGCC	32	30	10	0.521	$1.9 \cdot 10^{-5}$
8·10	GTATCGCC	29	23	12	0.475	$3.5 \cdot 10^{-5}$
8·12	GTAACCCC	30	23	11	0.564	$6.2 \cdot 10^{-5}$
8·16	ACCCAACC	29	22	12	0.602	$1.4 \cdot 10^{-4}$
8·20	AATGCAGC	27	18	10	0.655	$5.4 \cdot 10^{-4}$
8·23	CCGAGCTG	31	23	11	0.697	$4.1 \cdot 10^{-4}$
8·26	CCAGGTGG	14	9	10	0.747	$4.7 \cdot 10^{-2}$

[a] Number of amplification products of less than 500 bp.

[b] Number of unique fingerprint phenotypes identified by each primer in 13 cultivars.

[c] Estimated average level of band sharing ("true" average probability of a match between fingerprints of unrelated cultivars).

[d] Probability that a sample selected at random has a matching fingerprint, x^n, where n is the average number of loci per cultivar.

REFERENCES

Akkaya, M.S., A.A. Bhagwat, and P.B. Cregan. 1992. Length polymorphism of simple sequence repeat DNA in soybean. *Genetics,* 132:1131–1139.

Bassam, B. and S. Bentley. 1995. Arbitrary primer technology: basic theory and practice. In: *10th Biennial Australasian Plant Pathology Society Conference Workshop Manual*, pp. 16–25.

Bent, A.F., B.N. Kunkel, D. Dahlbeck, K.L. Brown, R. Schmidt, J. Giraudat, J. Leung, and B.J. Staskawicz. 1994. *RPS2* of *Arabidopsis thaliana*: A leucine-rich repeat class of plant disease resistance genes. *Science,* 265:1856–1860.

Caetano-Anollés, G. and P.M. Gresshoff. 1994a. Staining nucleic acids with silver: an alternative to radioisotopic and fluorescent labeling. *Promega Notes,* 45:13–18.

Caetano-Anollés, G. and P.M. Gresshoff. 1994b. DNA amplification fingerprinting using arbitrary mini-hairpin oligonucleotide primers. *Bio/Technology,* 12:619–623.

Caetano-Anollés, G. and P.M. Gresshoff. 1994c. DNA amplification fingerprinting of plant genomes. *Methods of Molecular and Cellular Biology,* 5:62–70.

Caetano-Anollés, G. and P.M. Gresshoff. 1996. Generation of sequence signatures from DNA amplification fingerprints with mini-hairpin and microsatellite primers. *Bio-Techniques,* 20:1044–1056.

Caetano-Anollés, G., B. Bassam, and P.M. Gresshoff. 1991. DNA amplification fingerprinting using very short arbitrary oligonucleotide primers. *Biotechnology,* 9:553–557.

Caetano-Anollés, G., L.M. Callahan, and P.M. Gresshoff. 1997. Inferring the origin of bermudagrass (*Cynodon*) off-types by DNA amplification fingerprinting in phytoforensic applications. *Crop Science,* 37:81–87.

Caetano-Anollés, G., L.M. Callahan, P.E. Williams, K. Weaver, and P.M. Gresshoff. 1995. DNA amplification fingerprinting analysis of bermudagrass (*Cynodon*): genetic relationships between species and interspecific crosses. *Theor. Appl. Genetics,* 91:228–235.

Gresshoff, P.M. 1993. Molecular genetic analysis of nodulation genes in soybean. *Plant Breeding Reviews,* 11:275–318.

Gresshoff, P.M. 1994. Plant genome analysis by single arbitrary primer amplification. *Probe,* 4:32–36.

Gresshoff, P.M. 1995a. Moving closer to the positional cloning of legume nodulation genes. In *Nitrogen Fixation: Horizons and Application.* Eds. Tikhonovich, I., N. Provorov, V. Romanov, and W.E. Newton, Kluwer Academic Publishers, Dordrecht, Netherlands, pp. 431–436.

Gresshoff, P.M. 1995b. The interface between RFLP techniques, DNA amplification and plant breeding. In *New Diagnostics in Crop Sciences*. Eds. J.H. Skerritt and R. Appels. CAB International, England. pp. 101–125.

Gresshoff, P.M. and A. MacKenzie. 1994. Low experimental variability of DNA profiles generated by arbitrary primer based amplification (DAF) of soybean. *Chinese J. Botany,* 6:1–6.

Landau-Ellis, D., S. Angermüller, R. Shoemaker, and P.M. Gresshoff. 1991. The genetic locus controlling supernodulation in soybean (*Glycine max* L.) co-segregates tightly with a cloned molecular marker. *Molecular and General Genetics,* 228:221–226.

Martin, G.B., S. Brommonschenkel, J. Chunwogse, A. Frary, M.W. Ganal, R. Spivey, T. Wu, E.D. Earle, and S.D. Tanksley. 1993. Map-based cloning of a protein kinase gene conferring disease resistance in tomato. *Science,* 262:1432–1436.

Mindrinos, M., F. Katagiri, G.-L. Yu, and F.M. Ausubel. 1994. The *A. thaliana* disease resistance gene *RPS2* encodes a protein containing a nucleotide-binding site and leucine-rich repeats. *Cell* 78:1089–1099.

Moore, G., M.D. Gale, N. Kurata, and R.B. Flavell. 1993. Molecular analysis of small grain legume genomes: status and prospects. *Bio/Technology,* 11:584–585.

Pillai, S., R.P. Funke, and P.M. Gresshoff. 1996. Yeast and bacterial artificial chromosomes (YAC and BAC) clones of the model legume *Lotus japonicus. Symbiosis,* 21:149–164.

Prabhu, R.R. 1995. *Soybean genome analysis using arbitrary primer technology.* PhD Dissertation, The University of Tennessee, Knoxville, TN.

Prabhu, R.R. and P.M. Gresshoff. 1994. Inheritance of polymorphic markers generated by short single oligonucleotides using DNA amplification fingerprinting in soybean. *Plant Molecular Biology,* 26:105–116.

Prabhu, R., H. Jessen, D. Webb, S. Luk, S. Smith, and P.M. Gresshoff. 1997. Genetic relatedness among soybean (*Glycine max* L.) lines revealed by DNA amplification fingerprinting, RFLP and pedigree data. *Crop Science,* In press.

Rongwen, J., M.S. Akkaya, A.A. Bhagwat, U. Lavi,. and P.B. Cregan. 1995. The use of microsatellite DNA markers for soybean genotype identification. *Theor. Applied Genetics,* 90:43–48.

Rommens, J.M., M.C. Iannuzzi, B.-S. Kerem, M.L. Drumm, G. Meimer, M. Dean, R. Rozmahel, J.L. Cole, D. Kennedy, N. Hidaka, M. Zsiga, M. Buchwald, J.R. Riordan, L.-C. Tsui, and F.S. Collins. 1989. Identification of the cystic fibrosis gene: chromosome walking and jumping. *Science,* 245:1059–1065.

Shizuya, H., B. Birren, U.-J. Kim, V. Mancino, T. Slepak, Y. Tachiiri, and M. Simon. 1992. Cloning and stable maintenance of 300-kilobase-pair fragments of human DNA in *Escherichia coli* using an F-factor-based vector. *Proc. Natl. Acad. Sciences* (USA), 89:8794–8797.

Song, W.Y., G.L. Wang, L. Chen, K.S. Kim, T. Holsten, B. Wang, Z. Zhai, L.H. Zhu, C. Fauceut, and P.C. Ronald. 1995. The rice disease resistance gene, Xa21, encoded receptor kinase-like protein. *Science,* 270:1804–1806.

Vos, P., R. Hogers, M. Bleeker, M. Reijans, T. van de Lee, M. Hornes, A. Frijters, J. Pot, J. Peleman, M. Kuiper, and M. Zabeau. 1995. AFLP: a new technique for DNA fingerprinting. *Nucleic Acid Research,* 23:4407–4414.

Weaver, K., G. Caetano-Anollés, P.M. Gresshoff, and L.M. Callahan. 1994. Isolation and cloning of DNA amplification products from silver stained polyacrylamide gels. *BioTechniques,* 16:226–227.

Weaver, K., L.M. Callahan, G. Caetano Anollés, and P.M. Gresshoff. 1995. DNA amplification fingerprinting and hybridization analysis of centipedegrass. *Crop Science,* 35:881–885.

Webb, D.M., B.M. Baltazar, A.P. Rao-Arelli, J. Schupp, K. Clayton, P. Keim, and W.D. Beavis. 1995. Genetic mapping of soybean cyst nematode race-3 resistance loci in the soybean PI 437.654. *Theor. Applied Genetics,* 91:574–581.

Weising, K., F. Weigand, A.J. Driesel, G. Kahl, H. Zischler, and J.T. Epplen. 1989. Polymorphic simple GATA/GACA repeats in plant genomes. *Nucleic Acids Research,* 17:10128–10131.

Weising, K., J. Kaemmer, F. Weigand, J.T. Epplen, and G. Kahl. 1992. Oligonucleotide fingerprinting reveals various probe-dependent levels of informativeness in chickpea (*Cicer aruentinum*). *Genome,* 35:436–442.

Welsh, J. and M. McClelland. 1990. Fingerprinting genomes using PCR with arbitrary primers. *Nucleic Acids Research,* 18:7213–7218.

Wicking, C. and B. Williamson. 1991. From linked marker to gene. *Trends in Genetics,* 7:288–290.

Williams, J.G.K., A.R. Kubelik, K.I. Livak, J.A. Rafalski, and S.V. Tingey. 1990. DNA polymorphisms amplified by arbitrary primers are useful as genetic markers. *Nucleic Acids Research,* 18:6531–6535.

Woo, S.S., V.K. Rastogi, H.-B. Zhang, A.H. Paterson, K.F. Schertz, and R.A. Wing. 1995. Isolation of megabase-size DNA from sorghum and applications for physical mapping and bacterial and yeast artificial chromosome library construction. *Plant Molecular Biology Reporter,* 13:82–94.

Young, N.D. and R.L. Phillips. 1994. Cloning plant genes known only by phenotype. *Plant Cell,* 6:1193–1195.

Zhu, T., L. Shi, R.P. Funke, P.M. Gresshoff, and P. Keim. 1996. Characterization and application of soybean YACs to molecular cytogenetics. *Mol. Gen. Genetics*, 252:483–488.

Chapter 2

Genetic Characterization of Open-Pollinated Turfgrass Cultivars

D.R. Huff

The ability to properly identify cultivars of turfgrass is vital to all sectors of the green industry. From the scientists to the producers to the end users, accurate cultivar identification (characterization) is becoming increasingly important. In this chapter, I will address several issues concerning the genetic characterization of cultivated populations of open-pollinated turfgrass species based on molecular markers. First, it is necessary to understand and appreciate the various forms of reproductive biology of grasses and how these reproductive modes are related to the process of cultivar development. Second, I will demonstrate how molecular data may be statistically analyzed for purposes of characterization. Third, I will cover the interpretation of results as related to sample size and the number and effect of individual molecular markers. Fourth, and finally, I will address the issue of genetic diversity as it relates to cultivar development, and introduce the notion of how molecular markers may enable golf courses to enhance the environment.

The turf industry utilizes over 30 species of grasses for turf (Table 2.1). To varying degrees, each of these species has a different evolutionary history, a different genomic organization, and a different reproductive strategy. Taken as a whole, these grasses range in variation from being: self-pollinated to cross-pollinated, apomictic to dioecious, diploids to octodecaploids, autoploids to alloploids, and from vegetatively propogated cultivars to open-pollinated seeded cultivars. While each of these genetic parameters is important to the breeder, the difference in cultivar development is vital to the following discussion of cultivar characterization.

Table 2.1. A Partial List of Important Turfgrass Species and Native Grasses Showing Their Respective Chromosome Number and Whether Typical Cultivars of Each Species are Either Homogeneous (Basically a Single Genotype) or Heterogeneous (a Mixture of Genotypes). *Note:* Species with Both Descriptors are Propagated Both Vegetatively (Homogeneous) and Through Seed (Heterogeneous).

Species	Number of Chromosomes	Genotypes Within a Cultivar or Population
***C_3 turfgrass species*:**		
Kentucky bluegrass	28 thru 126 (apomictic)	homogeneous (seeded)
Perennial ryegrass	14	heterogeneous
Tall fescue	42	heterogeneous
Creeping bentgrass	28	heterogeneous or homogeneous
Creeping red fescue	42, 56	heterogeneous
Sheep/Hard fescue	28/42	heterogeneous
Annual bluegrass	28 (self-pollinated)	homogeneous (seeded)
***C_4 turfgrass species*:**		
Common bermudagrass	36	heterogeneous
Hybrid bermudagrass	27 sterile	homogeneous
Japanese lawngrass	40	homogeneous
Manilagrass	40	homogeneous
Mascarenegrass	40	homogeneous
St. Augustinegrass	18	homogeneous or heterogeneous
Buffalograss (native)	20, 40, 60 (autoploid)	heterogeneous or homogeneous
***Native reclamation grasses*:**		
Little bluestem	40	heterogeneous
Sand bluestem	60	heterogeneous
Canada wild rye	28	homogeneous
Crinkled hairgrass	28	heterogeneous
Indiangrass	20	heterogeneous
Switchgrass	18, 36, 54, 72, 90, 108	heterogeneous

Before the cultivars within a turfgrass species may be characterized, it is important to understand that the different forms of reproductive biology have an important effect on the process of cultivar development. In general, there are two distinct classes of turfgrass cultivars: 1) those based on a single genotype (individual) and 2) those based on a population of genetically unique individuals. Turfgrass species whose cultivars are based on a single genotype include those that are capable of being vegetatively propogated (e.g., bermudagrass), or have a self-pollinated (e.g., annual bluegrass) or an apomictic (e.g., Kentucky bluegrass) breeding system. Such cultivars may actually contain more than a single genotype, but for the most part these cultivars are genetically uniform and are referred to as being homogeneous. Cultivars of turfgrass species that possess an outcrossing breeding system (e.g., perennial ryegrass, tall fescue, and creeping bentgrass) are populations of genetically unique individuals and are referred to as being heterogeneous, or a mixture of genotypes. If we were to use automobiles for an analogy, the former would represent car dealerships where each lot would have only a single make and model, with specific options, and would be the same color (homogeneous). The latter would represent car dealerships where each car on a lot is different in some way (different makes, different models, etc.) but all cars on the same lot would possess some feature of commonality — for example, all would be different shades of red on one lot versus different shades of blue on another. Currently, in perennial ryegrass, the newest experimentals under consideration for cultivar status are all dark green, dense, and dwarf. They are so similar in morphology that we can no longer distinguish them apart by simple inspection (T. Salt, pers. comm.).

In order to present a molecular approach to characterization, let's represent the differences between these two classes of cultivars in an alternative way. Figure 2.1 is an attempt to represent what different cultivars within each of these two classes might look like, genetically, after being theoretically developed from the same germplasm source. Genetic variation within the germplasm source is represented using eight different graphic patterns which we will refer to as eight distinct DNA "fingerprints." It is easy to see that those circles representing cultivars derived from vegetative propagation, or from apomictic or self-pollinated species, only contain one graphic pattern, and hence have a single DNA "fingerprint." Thus, for homogeneous cultivars, all genetic variation resides between cultivars. Circles that represent cultivars developed through outcrossing each contain the same six graphical patterns, and hence each cultivar is capable of producing all six DNA "fingerprints." Thus, for heterogeneous cultivars, most of the genetic variation resides within a cultivar. The important point here is that the different graphical patterns fill different proportions of the circles representing the heterogeneous cultivars. Thus, there is a frequency difference in the area covered by different graphical patterns between cultivars. The same is true for genetic variation among real cultivars of the heterogeneous class. In order to characterize heterogeneous cultivars, we will partition the molecular genetic variation into within and between cultivar variance components.

Molecular markers provide an enormous wealth of genetic data at the level of individual genotypes. Moreover, there is a variety of different molecular markers available to generate this genetic data. So many different types and classes of marker systems exist that it would take an additional chapter to cover all the pro's and con's of each. For illustrative purposes, the following

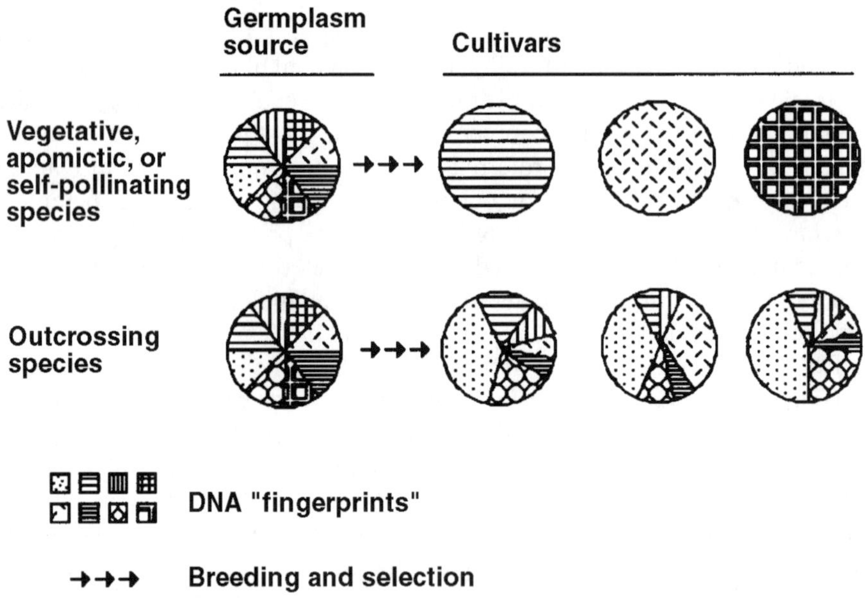

Figure 2.1. Hypothetical representation of genetic variation (DNA fingerprints) within and among cultivars of turfgrass species. Cultivars developed from vegetative propagation or from apomictic or self-pollinated species are homogeneous within cultivars, whereas cultivars developed from outcrossing (open-pollinated) species are heterogeneous within cultivars.

discussion will be based on results of mine using a class of molecular markers known as random amplified polymorphic DNA, or RAPDs; but the tools and techniques described are equally applicable to other types of markers as well. The genetic information generated by molecular markers is capable of being represented much like the graphical information was in Figure 2.1. Genetic relationships among individuals may be graphically represented using various clustering analyses (Figure 2.2a,b). Figure 2.2a depicts what is commonly known as a relatedness “tree,” whereas, Figure 2.2b uses the data’s first three eigenvectors to construct a genetic “space.” Notice that in Figure 2.2a, individuals 3 and 5 are separated from their cohorts (1, 2, 4, 6) and that individual 9 is likewise separated from its cohorts (7, 8, 10, 11, 12). In Figure 2.2b, however, all 24 individuals seem to correspond well to the four population classes (Mexico A 1-6; Mexico B 13-18; Texas A 7-12; and Texas B 19-24). These two methods of clustering analysis are simply different ways of visually presenting the same genetic relationships among all 24 individuals using either two- or three-dimensional space. The regional differences among these plants is easily distinguished using either analysis. However, in my experience, if a three-dimensional representation is justified according to the amount of total

(A) **A 2-dimensional UPGMA tree**

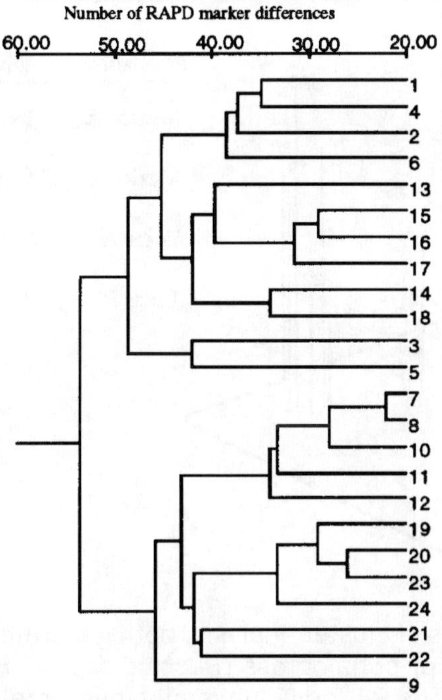

Populations	Individuals
Mexico A	1 - 6
Mexico B	13 - 18
Texas A	7 - 12
Texas B	19 - 24

Figure 2.2a. Different methods of cluster analysis, both generated using NTSYS-pc, for 24 individuals of buffalograss (*Buchloë dactyloides*) surveyed for 97 RAPD markers. The 24 individual genotypes were derived plants originally collected from four areas: two locations in Mexico (A and B) and two locations in Texas (A and B). The A locations are two-dimensional phenogram trees generated through the unweighted pair group clustering method using arithmetic averages. The B locations are three-dimensional representations of the first three eigenvectors from a principle coordinates analysis.

variation contained therein, it seems to more accurately display known genetic relatedness among individuals. Notice, too, that neither clustering analysis provides a statistic that would indicate if the four populations are statistically different from one another. There are methods for testing goodness-of-fit of the data to the graphical representation, but there is no test to reveal whether or not the four populations are statistically different. Of course, where there is no overlap between two populations, then we can assume them to be different (e.g., Mexico A versus either Texas population). However, when the individuals of two populations slightly overlap (as in Figure 2.2a), the question arises: Are these populations distinct enough to be considered different populations? If your answer is yes, then the real question becomes: How much overlap can be tolerated and still retain statistically different population structure? This

(B) A 3-dimensional principle coordinates analysis

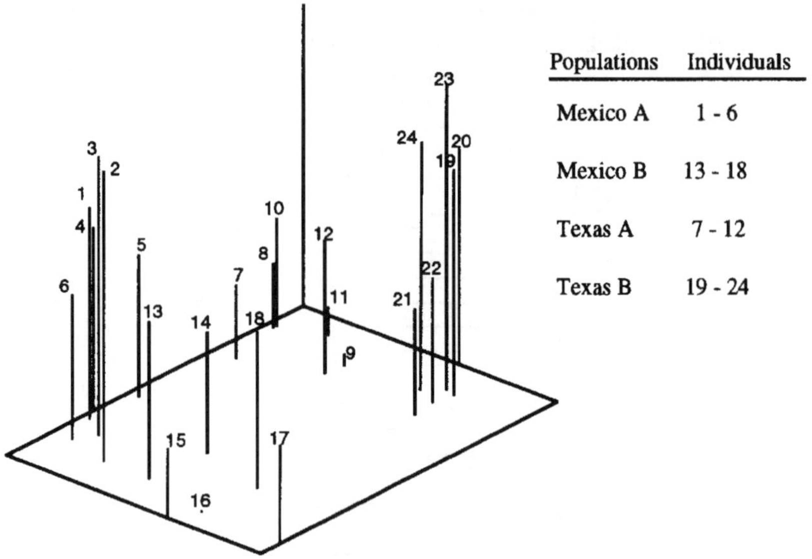

Figure 2.2b. Different methods of cluster analysis, both generated using NTSYS-pc, for 24 individuals of buffalograss (*Buchloë dactyloides*) surveyed for 97 RAPD markers. The 24 individual genotypes were derived plants originally collected from four areas: two locations in Mexico (A and B) and two locations in Texas (A and B). The A locations are two-dimensional phenogram trees generated through the unweighted pair group clustering method using arithmetic averages. The B locations are three-dimensional representations of the first three eigenvectors from a principle coordinates analysis.

last question is one we are currently addressing for characterizing cultivars of some of the most important turfgrass species.

Fortunately, there is a statistical tool that helps us to test differences between overlapping populations. The statistical tool we employ is known as analysis of molecular variance (AMOVA). Rather than using graphical symbols to represent genetic differences among individuals, we use a parameter known as genetic distance, which is calculated directly from the molecular marker data. There are many different methods for calculating genetic distance. The form of genetic distance used by AMOVA is known as the Euclidean metric distance. It's referred to as Euclidean because it conforms to the 300 B.C. Greek geometer Euclid's axioms and definitions of geometric space. By conforming to geometric space, the Euclidean metric distance allows AMOVA to partition 100% of the total genetic variation into component sources of variation. Thus, in this way, AMOVA resembles something familiar: a standard nested analysis of variance (Table 2.2, i). For testing purposes, the vari-

Table 2.2. Structure of Analysis of Molecular Variance (AMOVA) Output.

i) Global nested model:

Source	df	SS	MS	% Total Variation
Regions	df (A)	SS (A)	MS (A)	V (A)
Populations within regions	df (B)	SS (B)	MS (B)	V (B)
Individuals within populations	df (C)	SS (C)	MS (C)	V (C)
Total	—	—	—	V (T) = 100%

ii) Variance components significance testing (permutational techniques):

V (A) and PHIct
V (B) and PHIsc
V (C) and PHIst

iii) Distances among populations:

where: distances = PHIst between pairs of populations (below diagonal).

and, distances between populations are tested using permutational testing techniques involving 1,000 permutations (above diagonal).

Example:

		Populations		
1	**2**	**3**	**4**	**5**
—	0.005	0.03	0.001	0.001
0.06	—	0.007	0.13	0.001
0.04	0.06	—	0.05	0.001
0.09	0.03	0.04	—	0.001
0.15	0.12	0.10	0.16	—

ance components are equivalent to parameters known as PHI statistics (Table 2.2, ii). In AMOVA, these PHI statistics are analogous to "F-statistics" used in the genetic analysis of gene frequencies. AMOVA also generates a PHIst statistic between pairwise comparisons of populations (Table 2.2, iii). PHIst is equivalent to the percentage of the variation that is partitioned between any two populations, and may be used as a measure of the genetic distance between populations. In the example provided, all populations are significantly different ($p \leq 0.05$) from each other, with the exception of populations 2 and 4.

Although AMOVA resembles an analysis of variance, it does not determine significance of variance components by comparing distributions against the error term. Rather, significance testing is performed using a permutational technique. To illustrate how this permutational technique operates, let's consider sets of hands in the card game known as bridge (Figure 2.3). For this demonstration, we will consider the cards to represent the individual turfgrass genotypes, and the players (north, south, east, and west) to represent four cultivars (or populations). Permutational testing procedures take the cards from the observed hand, shuffle them, and then redeal all the cards out to the four players. Variation within and among players is then recalculated for each of the hands. The between-player variance component of the computer-generated hands is then compared to that of the observed hand variance. If the computer is capable of generating larger between-population (player) variance components through random shuffling, say 50% of the time, than the one that was calculated from the observed hand, then the observed variance estimate is not very meaningful. In other words, the population structure of the observed hand could have easily happened by chance. If, however, random permutation of the data does not produce a variance component larger than the observed, then the observed population structure exceeds random expectation and is considered to be significant. Varying degrees of significance are the result of the computer being able to occasionally generate a larger than observed variance, say 4 times out of 100, versus 2 times out of 100, versus 1 time out of 1,000.

DISCUSSION

Cultivars of turfgrasses are developed using various breeding methodologies, including backcrossing, open-pollinated bulk harvests from maternal parents, and even natural ecotype selections. Depending on the relatedness of the germplasm involved, the resulting cultivars may represent a range of genetic distances. The statistical tools and techniques presented offer a means to detect and characterize genetic distances among heterogeneous cultivars. In addition, these techniques may be useful for examining the effects of genetic drift in the production of these cultivars over time and space.

Some general observations of my research suggest that when two populations overlap, our ability to detect population differences is determined by three items. (1) *Inherent genetic distance.* The detectable limits of population significance, via AMOVA, seem to be when approximately 5% of the variation resides between populations. Populations closer than PHIst = 0.05 tend to be nonsignificantly different under the permutational testing methods. As PHIst values exceed 0.05, significance becomes more easily detectable. (2) *Sample size.* When it comes to sample size, more is always better, but there are usually diminishing returns after a certain point. Genetic distance between populations includes a wide range of possible values. When populations are extremely distant, even small sample sizes (as low as two individuals) are sufficient to show population difference. Populations that are closer in distance will require a larger sample size in order to detect significant differences. For initial surveys within a species, I usually start with sample sizes of 10 to 12 per population. This size seems to yield fairly precise estimates of genetic distance — at least precise enough to base further testing upon. Populations that are found to be extremely close may require larger sample sizes for more pre-

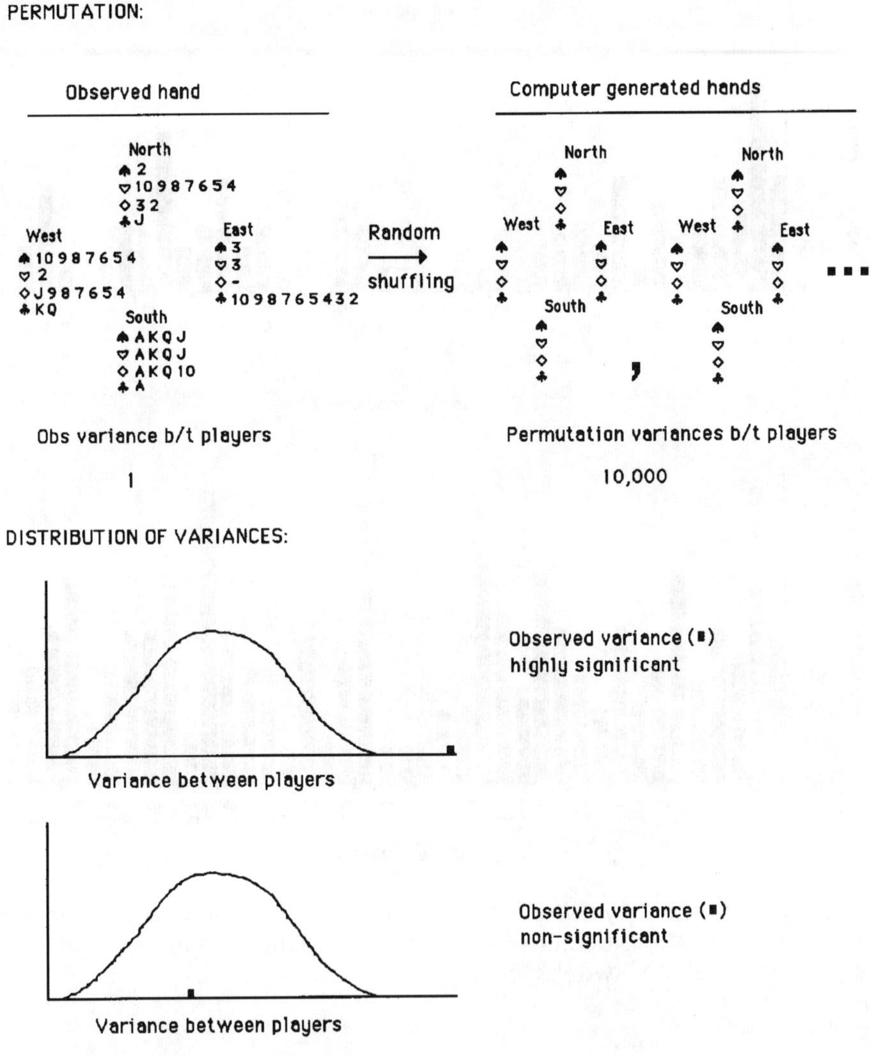

Figure 2.3. A hypothetical representation of permutational testing techniques. Permutation was performed on playing cards from the game known as bridge. The distribution of between-player variances of the computer-generated permutations is compared to the observed between-player variance from the observed hand.

cise estimates of distance. (3) *Effect of individual markers.* As the number of markers surveyed increases, the amount of information also increases, and the chances of detecting population differences become greater. However, the discriminatory power of some molecular markers is greater than that of other markers (Figure 2.4). The ability of an individual marker to discriminate between two populations is related to its frequency difference between those

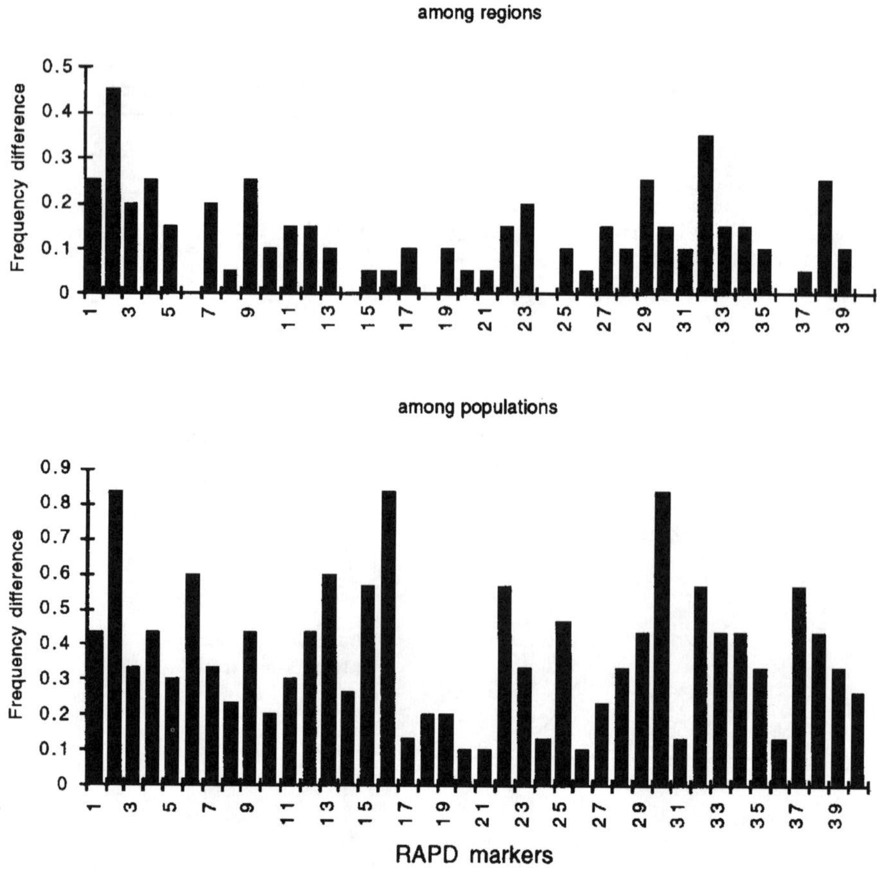

Figure 2.4. Frequency differences for 40 RAPD markers observed from actual data of little bluestem (*Andropogon scoparius*) collected from two populations each of two regions.

populations. Thus, some markers may have little to no discriminatory power, even though they are polymorphic (variable) within populations. For example, imagine a marker that is present in 50% of the individuals for every population examined. Such a marker provides no discrimination among populations. Thus, indiscriminate use of a large number of markers is not necessarily an efficient use of time and supplies compared to screening for a few highly discriminating markers.

Given a large sample size and a large number of informative markers, it is possible to imagine a situation where there is still sufficient overlap of genotypes to prevent separation of populations. In such a situation, we would be unable to reject the null hypothesis that the individual genotypes are from the same population. In this situation, genetic diversity becomes an important issue for guiding plant breeding decisions on the selection of germplasm resources. From knowing the breeding histories and response to selection for some of our turf species, there is cause for concern that we may have reached a dangerously low level of genetic diversity among our cultivars and germplasm sources. The tools and techniques presented would be a useful way of measuring and monitoring genetic diversity among our germplasm sources.

The issue of genetic diversity is also a central issue concerning the biodiversity of our environment. There is a large movement in practice today to restore native ecosystems, thereby stabilizing and enhancing our planet's biodiversity at local levels. This movement can even be seen on golf course roughs and nonplay areas, where grasses are being allowed to grow to their mature height to provide wildlife habitats. Currently, U.S. military bases are considered to be our greatest asset for maintaining biodiversity. I believe golf courses could become one of our greatest assets for preserving and maintaining genetic diversity of our native grass species. What we will need to avoid, in the future, is the replacement and/or the genetic contamination of our local native grass populations with standard commercial types that may be genetically very different from endemic local populations. This is because a native grass population is more than just a Latin binomial. It implies that evolutionary forces may have created something unique and irreplaceable at the local level. The same techniques used for characterizing heterogeneous, open-pollinated cultivars of introduced turfgrass species are also capable of preserving our local biodiversity by preventing their genetic contamination by "foreign" sources of native grasses. Whether or not commercial sources of native grass species pose a genetic threat to local native grass populations is a major area of focus in my current research.

REFERENCES

Excoffier, L., P.E. Smouse, and J.M. Quattro. 1992. Analysis of molecular variance inferred from metric distances among DNA haplotypes: application to human mitochondrial DNA restriction data. *Genetics,* 131:479–491. (Program available from L. Excoffier @ tel: 41-22-702 69 65; E-mail: excoffie@sc2a.unige.ch).

Funk, C.R., J. Murphy, and D.R. Huff. 1993. Diversity and vulnerability of perennial ryegrass, tall fescue, and Kentucky bluegrass. Turfgrass Germplasm Symposium. *Agronomy Abstracts.* p. 188.

Huff, D.R. 1997. RAPD characterization of heterogeneous perennial ryegrass cultivars. *Crop Science.* 37:557–564.

Huff, D.R. and J.M. Bara. 1993. Determining genetic origins of aberrant progeny from facultative apomictic Kentucky bluegrass using a combination of flow cytometry and silver-stained RAPD markers. *Theor. Appl. Genet.,* 87:201–208.

Huff, D.R., R. Peakall, and P.E. Smouse. 1993. RAPD variation within and among natural populations of outcrossing buffalograss [*Buchloë dactyloides* (Nutt) Engelm.]. *Theor. Appl. Genet.,* 86:927–934.

Knapp, E.E. and K.J. Rice. 1996. Genetic structure and gene flow in *Elymus glaucus* (blue wildrye): implications for native grassland restoration. *Restoration Ecology,* 4:1–10.

Peakall, R., P.E. Smouse, and D.R. Huff. 1995. Evolutionary implications of allozyme and RAPD variation in diploid populations of dioecious buffalograss [*Buchloë dactyloides* (Nutt.) Engelm.]. *Molecular Ecology,* 4:135–147.

Rohlf, F.J. 1989. NTSYS-pc: Numerical Taxonomy and Multivariate Analysis System. Exter Publishers, Setauket, NY. (Cluster analyses program available from Applied Biostatistics @ (516)751–1203).

Salt, T. 1996. Plant Variety Protection Office, AMS, Beltsville, MD.

Chapter 3

Isozyme Genetics, and Cultivar Relationships of Creeping Bentgrass (*Agrostis palustris* Huds.)

S.E. Warnke, D.S Douches, and B.E. Branham

INTRODUCTION

The bentgrasses are native to Western Europe (Harlan, 1992) with the genus *Agrostis* consisting of approximately 200 species (Hitchcock, 1951). The four species commonly accepted as turfgrass types are *Agrostis palustris* Huds. — creeping bentgrass (2n=4x=28), *Agrostis canina* L. — velvet bentgrass (2n=2x=14), *Agrostis tenuis* Sibth. — colonial bentgrass (2n=4x=28), and *Agrostis alba* L. — redtop bentgrass (2n=6x=42). Because of its excellent tolerance of low mowing heights and a strong stoloniferous growth habit, creeping bentgrass is well suited for the establishment of golf greens and fairways, and is thus the most widely utilized of the turf-type bentgrasses.

GENETIC STUDIES

Limited genetic research has been conducted on the genus *Agrostis*. Jones (1956) examined chromosome pairing in tetraploid *A. tenuis* Sibth. and *A. stolonifera* L. and concluded that both *A. tenuis* and *A. stolonifera* form bivalents at meiosis. *A. tenuis* behaved as a segmental allotetraploid and *A. stolonifera* as a strict allotetraploid with well-differentiated genomes. However, chromosome pairing studies are not a definitive method for discriminating between allo- and autopolyploid inheritance because autotetraploid inheritance can still occur through random bivalent pairing (Krebs and Hancock, 1989).

Genetic data from codominant molecular markers can clearly distinguish between allo- and autopolyploidy due to a difference in the progeny classes observed from crosses. A diploid organism with two different alleles at a locus (*A* and *a*) produces one heterozygote class (*Aa*). Self-fertilization of this individual will produce a progeny array of 1*AA*:2*Aa*:1*aa*. However, with tetrasomic inheritance, in an autotetraploid, three different classes of heterozygotes can be produced (*AAaa*), (*Aaaa*) and (*AAAa*). Self-fertilization of an individual having the (*AAaa*) genotype will result in a progeny array of 1*AAAA*:8*AAAa*:18*AAaa*:8*Aaaa*:1*aaaa* (Muller, 1914; Haldane, 1930). In contrast, preferential pairing of genomes in an allotetraploid having *AA* on a chromosome pair in one genome and *aa* on a chromosome pair from the other genome would produce only AAaa progeny. Therefore, allopolyploidy can result in "fixed" heterozygosity, which would not be expected in a autopolyploid.

Isozymes are a class of codominant molecular markers that have been useful for the study of polyploid inheritance in a number of plant species. Therefore, the objectives of our research efforts were to study inheritance in cultivated creeping bentgrass using isozymes and to estimate the genetic diversity of cultivated creeping bentgrasses based on isozyme polymorphisms.

Progeny for isozyme analysis were developed from crosses among highly fertile bentgrass clones selected for their differences in isozyme banding patterns. The crossing of selected clones was attempted by three different methods: (1) All clones except Syn91-3 were placed in pollination isolation. (2) Hot water emasculation was attempted by submerging one clone of a cross in hot water. (3) Reproductive tillers from field-grown plants were placed in 30-mL test tubes held upright in clay pots. Clear plastic bags supported by wooden stakes were placed over the tillers to maintain pollination isolation. The pots were placed in the greenhouse and the test tubes filled with water as needed. Seed harvested from the two parents of each cross was kept separate to examine whether any self-fertilization had occurred. Chi-square tests were used to test hypothesized genetic ratios. Whenever possible, plants homozygous for alternate alleles at a locus were selected for crossing to determine if self-fertility occurred. Plants appearing to exhibit a duplex allelic state (i.e., AABB) at a putative locus were also selected, since inheritance data from these plants provides the clearest differentiation between disomic and tetrasomic inheritance in self-fertilizing and testcross progenies.

The segregation data from crosses that exhibit fixed heterozygosity as well as the tetra-allelic segregation, observed for the cross of PE 91-25 with NA 91-22 and PV 91-21, provides strong genetic evidence for disomic rather than tetrasomic inheritance in creeping bentgrass (Table 3.1). Tetrasomic inheritance, either by the formation of quadrivalents or random bivalent pairing, would

Table 3.1. Segregation and Chi-Square Values at 5 Polymorphic Loci in Creeping Bentgrass.

Locus	Cross	Cross Type	Observed Progeny Phenotypes								Ratio Tested	(χ^2)	P
				dd[a]	dg	df	dfg						
Pgi-2	PV 91-21 X NA 91-22	6			63	70	58	65			1:1:1:1	(2.74)	0.46
			bd	bdg	bde	bdeg	cd	cdg	cde	cdeg			
	PE 91-25 X NA 91-22	11	19	18	24	21	20	19	22	18	1:1:1:1:1:1:1:1	(1.52)	0.98
			bd	bdf	bde	bdef	cd	cdf	cde	cdef			
	PE 91-25 X PV 91-21	10	13	23	11	17	18	24	12	14	1:1:1:1:1:1:1:1	(10.2)	0.23
						dd	df						
	Nor 91-7 X SR 91-30	4				85	106				1:1	(2.31)	0.17
						ab	abc						
Tpi-2	PE 91-25 X PV 91-21	7				73	74				1:1	(0.01)	0.95
	PV 91-21 X PE 91-24	7				120	115				1:1	(0.11)	0.81
	Nor 91-7 X SR 91-30	5				192					0:1:0	(0.00)	1.00
Got-2	PE 91-25 X NA 91-22	7				77	68				1:1	(0.56)	0.48
	PE 91-25 X PV 91-21	8				172					1:0	(0.00)	1.00
						aa	ac						
	Syn 91-3 ⊗	2				20	58				1:3	(0.13)	0.89
						bb	bc						
Pgm-2	Nor 91-7 X SR 91-30	1				100	87				1:1	(0.90)	0.42
						cc	bc						
Tpi-1	Syn 91-3 ⊗	9				9	34				1:3	(0.31)	0.57
	PV 91-21 X PE 91-24	3				205					0:1	(0.00)	1.00

[a] **Pgi-2**: b=*Pgi-2*^{2}, c=Pgi2^3, d=*Pgi2*^{4}, e=*Pgi2*^{5}, f=*Pgi-2*^{6}, g=*Pgi-2*^{7}; **Tpi-1**: b=*Tpi-1*^{2} c=*Tpi-1*^{3}; **Tpi-2**: a=*Tpi-2*^{1}, b=*Tpi-2*^{2}, c=*Tpi-2*^{3}; **Got-2**: a=*Got-2*^{1}, b=*Got-2*^{2}, c=*Got-2*^{3}; **Pgm-2**: b=*Pgm-2*^{2}, c=*Pgm-2*^{3}.

result in 12 different progeny classes, rather than the 8 that are observed in these crosses and expected with disomic inheritance.

GENETIC DIVERSITY

An understanding of the genetic diversity present in the cultivated germplasm of a species, as well as the location of new sources of genetic variability, is important for the optimal utilization of genetic resources. Plant breeders must have an understanding of the genetic variability of elite germplasm because continued reselection within this germplasm can narrow the genetic base of elite material and ultimately increase the potential vulnerability to pests and abiotic stresses. Information about the location of new sources of genetic variability can help broaden the genetic base of elite material and maintain long-term improvement.

Information about the relatedness of creeping bentgrass cultivars is limited because it is an allogamous species and in many cases the parental clones used in synthetic cultivar development are of unknown origin, making accurate estimates of relatedness based on coefficients of parentage impossible. However, in a number of species, estimates of genetic similarity between cultivars have been determined based on molecular markers such as isozymes. Isozyme markers are not as numerous as RFLP or RAPD markers; however, they are polymorphic in creeping bentgrass populations (Warnke et al., 1997b) and technically simpler than RFLP or RAPD markers when dealing with large population sizes.

Eighteen *Agrostis palustris* Huds. cultivars representing most of the named creeping bentgrass cultivars commercially available in the United States and one plant introduction were assayed for isozyme polymorphisms. Seed was obtained from the 1993 National Turfgrass Evaluation Program's (NTEP) bentgrass cultivar trial or directly from the seed company that produced the cultivar. In a few cases seed from the 1989 NTEP bentgrass cultivar trial was used (Table 3.1).

Each gel contained 25 plants from a cultivar and 2 check plants, of known allozyme composition, to aid in band scoring. Allelic bands (Warnke et al., 1997b) were scored as 1 (present) or 0 (absent) for each plant. A total of 75 plants from each cultivar were analyzed. However, due to enzyme degradation in some samples, fewer plants were used to establish band frequencies. The band frequency in the population was obtained by dividing the number of plants containing the band by the total plants examined. Genetic distances between and within each cultivar were established using Nei's 1972 distance formula. A dendrogram based on the distance matrix was constructed by applying the unweighted pair group method with arithmetic averages (UPGMA) cluster analysis. The distance matrix and dendrogram were both constructed using NTSYS-pc version 1.7 (Rohlf, 1992). Between-variety genetic distances were calculated using sample sizes of 4, 8, 12, 16, 20, 25, and 70 to establish

Table 3.2. Creeping Bentgrass Cultivars Studied, the Number of Plants Scored in Each, and the Average Within-Cultivar Genetic Distance (AWGD) of Each.

Cultivar	Year[a]	Source	Sponsor	Number Scored	AWGD[b]
1 Penneagle	1978	1993 NTEP bentgrass trial	Tee-2-Green Corp.	72	0.233
2 Penncross	1955	1993 NTEP bentgrass trial	Tee-2-Green Corp.	73	0.210
3 Trueline	1995	1993 NTEP bentgrass trial	Turf Merchants	73	0.267
4 Crenshaw	1993	1993 NTEP bentgrass trial	Loft's Seed	73	0.203
5 Southshore	1992	1993 NTEP bentgrass trial	Loft's Seed	73	0.326
6 Providence	1988	1993 NTEP bentgrass trial	Seed Research	70	0.230
7 National	1988	1989 NTEP bentgrass trial	Pickseed West	70	0.312
8 Viper	1995	International Seeds	International Seeds	72	0.173
9 18th Green	1995	1993 NTEP bentgrass trial	Johnson Seeds	73	0.257
10 Cobra	1988	1989 NTEP bentgrass trial	International Seeds	73	0.180
11 Emerald	1973	1989 NTEP bentgrass trial	International Seeds	73	0.138
12 Pennlinks	1987	1993 NTEP bentgrass trial	Tee-2-Green Corp.	73	0.387
13 SR1020	1987	Seed Research	Seed Research	73	0.265
14 Putter	1988	Jacklin Seed	Jacklin Seed	73	0.161
15 Seaside	1924	1993 NTEP bentgrass trial	Standard	73	0.258
16 Lopez	1994	1993 NTEP bentgrass trial	Finelawn Research	72	0.325
17 Pro/Cup	1994	1993 NTEP bentgrass trial	Forbes Seed Grain	73	0.340
18 Cato	1993	1993 NTEP bentgrass trial	Pickseed West	70	0.152
19 PI251945	1958[c]	Plant Intro. Station, Pullman, WA		25	0.104

[a] Year of release.
[b] Average within-cultivar genetic distance calculated using Nei's 1972 distance formula.
[c] Year collected.

the optimum population sample size for estimating the genetic distance between varieties.

Distance values ranged from 0.007 for Southshore and Pennlinks to 0.277 for Cato and PI251945. The dendrogram resulting from the UPGMA cluster analysis is shown in Figure 3.1. The UPGMA cluster analysis separates the cultivars into two main groups. The first group includes 10 cultivars (Penneagle, Putter, Penncross, Trueline, Viper, Emerald, 18th Green, Cobra, Crenshaw, and Seaside). With the exception of Crenshaw these are strongly creeping cultivars having a prostrate to semi-erect growth habit. The cultivar Seaside is at the base of this group and may have provided much of the initial germplasm for creeping bentgrass cultivar development over the last 40 years. Seaside originated as a naturalized population growing in tidal flats near Coos Bay, Oregon and was the only widely available seeded bentgrass in the U.S. from the 1920s until 1955 when Penncross was released (Duich, 1985). The second cluster contains nine cultivars that can be divided into two groups. Four of these cultivars (Southshore, Pennlinks, Pro/Cup, and Lopez) cluster quite closely and are difficult to distinguish from one another with the isozyme polymorphisms studied. One reason for the tight clustering of these four cultivars is that they possess most of the allozymes observed in this study. This isozyme diversity is evident in the fact that these four cultivars have the highest average within-cultivar genetic distances (AWGD) of any of the cultivars tested. The cultivar Southshore is a very broad-based cultivar derived from the progenies of 203 selected clones (Hurley et al., 1994), which would explain its high AWGD. The last four cultivars in the second group (Providence, SR1020, National, and Cato) do not group closely with any of the other cultivars and, in fact, these cultivars all possess some unique isozyme characteristics. The plant introduction PI251945 was included in the study to gain some insight into the European bentgrass germplasm. PI251945 was collected in 1958 from Austria and, based upon these results, it is quite distantly related to the U.S. cultivars examined, suggesting that European germplasm may be a means to broaden the genetic diversity of U.S. germplasm.

CONCLUSIONS

The results from the isozyme genetics studies indicate that creeping bentgrass is behaving as a strict allotetraploid, as suggested by Jones in 1956. The combination of high levels of outcrossing and allotetraploidy combine to make creeping bentgrass a very polymorphic species. High levels of polymorphism will aid in the construction of linkage maps for marker assisted selection. The most efficient mapping strategy for use with creeping bentgrass would be double test-cross populations in which segregating markers from both parents of a cross are scored in segregating progeny. The double pseudotest-cross strategy (Weeden, 1994) allows for the construction of a map for each parent and has been used with other outcrossing species, such as apple.

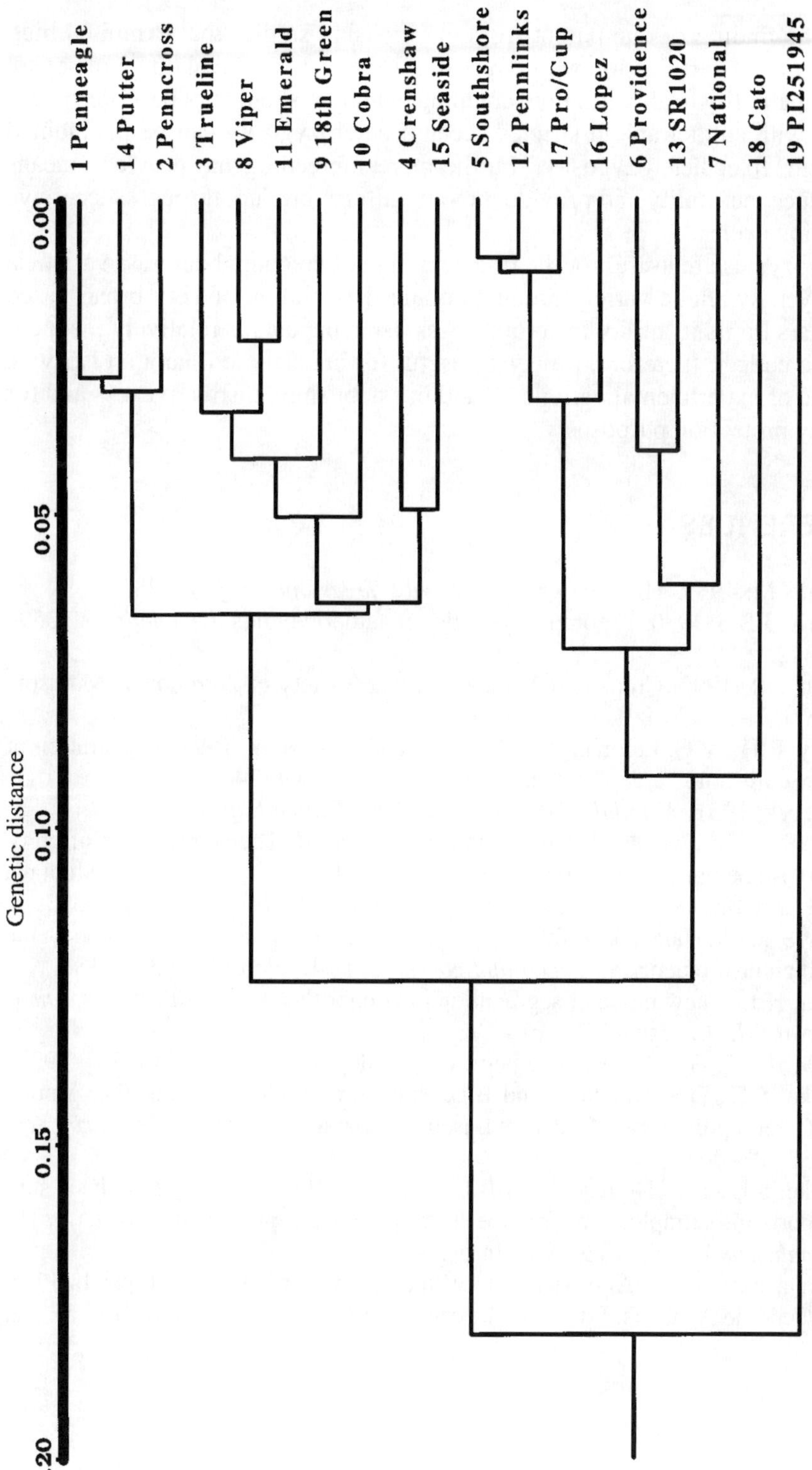

Figure 3.1. Dendrogram of 18 creeping bentgrass cultivars generated by UPGMA cluster analysis.

The finding of one parent from our crossing studies that exhibited high levels of self-fertilization was surprising. However, the breakdown of sexual incompatibility in allotetraploid creeping bentgrass should not be as detrimental as with autotetraploid species because heterozygosity can be maintained through "fixed heterozygosity." Furthermore, inbreeding may provide a means to reduce variability and provide a more uniform product through successive breeding cycles.

Isozymes are useful for the fingerprinting of creeping bentgrass varieties; however, synthetic variety production and the creation of very broad based cultivars limits its utility in certain cases. Isozyme data is relatively inexpensive to collect; therefore, it may be useful for breeders to obtain an isozyme profile of experimental varieties to aid in establishing distinctiveness and for variety protection purposes.

REFERENCES

Duich, J.M. 1985. The bent grasses. *Weeds, Trees and Turf,* 72–78.

Haldane, J.B.S. 1930. Theoretical genetics of autopolyploids. *J. Genetics,* 22:359–372.

Harlan, J.R. 1992. Crops and Man. American Society of Agronomy, Madison, WI.

Hurley, R.H., V.G. Lehman, J.A. Murphy, and C.R. Funk. 1994. Registration of "Southshore" creeping bentgrass. *Crop Sci.,* 34:1124–1125.

Hitchcock. 1951. *Manual of the Grasses of the United States.*

Jones, K. 1956. Species differentiation in *Agrostis* II. The significance of chromosome pairing in the tetraploid hybrids of *Agrostis canina* subsp. Montana Hartm., *A. tenuis* Sibth. and *A. stolonifera* L. *J. Genetics,* 54:377–393.

Krebs, S. and J. Hancock. 1989. Tetrasomic inheritance of isozyme markers in the highbush blueberry, *Vaccinium corymbosum* L. *Heredity,* 63:11–18.

Muller, H.J. A new mode of segregation in Gregory's tetraploid Primulas. *Amer. Nat. XLVIII,* 508–512, 1914.

Nei, M. 1972. Genetic distance between populations. *Am. Nat.* 106:283–292.

Warnke, S.E., D.S. Douches, and B.E. Branham. 1997a. Relationships Among Creeping Bentgrass Cultivars Based on Isozyme Polymorphisms. *Crop Sci.,* 37:203–207.

Warnke, S.E., D.S. Douches, and B.E. Branham. 1997b. Isozyme analysis supports allotetraploid inheritance in tetraploid creeping bentgrass. (*Agrostis palustris* Huds.). *Crop Sci.,* In press.

Weeden, N.F. 1994. Approaches to Mapping in Horticultural Crops: In Plant Genome Analysis. Ed. P.M. Gresshoff. CRC Press, Boca Raton, FL, 57–68.

Chapter 4

DNA Typing (Profiling) of Seashore Paspalum (*Paspalum vaginatum* Swartz) Ecotypes and Cultivars

S.M. Brown, S.E. Mitchell, C.A. Jester, Z.W. Liu,
S. Kresovich, and R.R. Duncan

ABSTRACT

With the increased interest and use of plant variety protection and the increased concern for protecting intellectual property rights for breeders and farmers, a highly discriminating and robust method based on DNA analysis is needed for the identification and characterization of turfgrass ecotypes and cultivars. Simple sequence repeats (SSRs), also known as microsatellites or short tandem repeats, are highly variable DNA sequences that have been proposed for use in identification and genetic analysis of plant and animal species. The use of semiautomated fluorescent detection of polymerase chain reaction (PCR) products has greatly improved the accuracy, speed, and cost of SSR typing. Five pairs of fluorescently labeled SSR-specific primers were employed to generate DNA profiles for 10 seashore paspalum (*Paspalum vaginatum* Swartz) accessions. Estimated gene diversity values for the five SSRs averaged 0.79 with a range from 0.74–0.87. These five molecular markers then were used to establish seven unique DNA profiles for the test array. Interestingly, "Excalibur," "Fidalayel," and FR-1 were found to be genetically identical to each other and approximately 85% similar to "Adalayd" (from which these three selections were made). The likelihood that the three selections shared the same DNA profile by chance was less than 1 in 500,000. This preliminary effort illustrates the value of applying SSRs for the cultivar identification and the determination of genetic relationships among seashore paspalum accessions.

INTRODUCTION

Seashore paspalum (*Paspalum vaginatum* Swartz) is an environmentally compatible, warm season grass that is being developed as a recreational and landscape turf and for the bioremediation of contaminated soil and water. The species requires minimal fertility and pesticides, and is adapted to several abiotic environmental stresses (drought, excessive water, salinity, acidity, and compacted or high bulk density soils). As seashore paspalum is improved through breeding, the need for — and value of — molecular genetic markers for cultivar/ecotype identification and the measurement of genetic variation also increases. In response, DNA marker technologies are becoming more widely adopted for both cultivar protection and as a tool for breeding program enhancement in crop and animal agriculture (Smith and Smith, 1992).

Over the past decade, molecular markers have been used to identify and characterize materials in a wide variety of crop species, ranging from field (Smith and Smith, 1992; Siedler et al., 1994; Rongwen et al., 1995; Virk et al., 1995) through horticultural crops (Yang and Quiros 1993; Graham et al., 1994; Novy et al., 1994; Tancred et al., 1994; Thomas et al., 1994; Yamagishi 1995). Specifically, molecular markers have been applied to the study of genetic variation in a number of turfgrass species, including buffalograss [*Buchloë dactyloides* (Nutt) Engelm] (Huff et al., 1993; Wu and Lin, 1994), centipedegrass [*Eremochloa ophiuroides* (Monro) Hackel] (Weaver et al., 1995), bermudagrass [*Cynodon* spp.] (Caetano-Anollés et al., 1995), and seashore paspalum (Liu et al., 1994 and 1995).

When considering the application of molecular markers to cultivar identification, both technical and operational issues must be considered. For example, technical issues relevant to marker characteristics include discriminatory ability, sensitivity, reproducibility, and the ability to be used for further genetic analysis or in diagnostics. Operational issues include protocol characteristics, time, and cost. The ideal molecular marker must be easy to employ, timely, cost effective, highly informative, and reliable (accurate with the desired level of precision). In order to be easy to employ, timely, and cost effective, sample preparation must be simple and the assay (including data generation, collection, organization, and analysis) should be suitable for increased throughput and automation. A high information content necessitates a marker assay that detects high heterozygosity, provides discriminatory ability among closely related individuals, and generates data from multiple genomic sites using a single assay. Reliability implies reproducibility of results from assay to assay both within and across laboratories, as well as unambiguous data analysis. In addition, a codominant marker would be advantageous for investigations involving mapping, pedigree analysis, and other plant breeding applications.

To date, many obstacles, including applicability, accuracy, reproducibility, and cost, have precluded the widespread adoption of DNA-based markers for plant variety protection. At present, markers based on the polymerase chain reaction (PCR) are receiving the most attention from researchers (Caetano-Anollés, 1994; Brown et al., 1995; Morell et al., 1995) because they have the potential for widespread, low-cost, large-scale application suitable for the multiple needs of plant breeding, genetic resources conservation, and cultivar profiling. A PCR-based assay requires only small amounts of crude genomic DNA preparations from each sample; the procedure is not technically challenging or expensive; and accurate results can be obtained in a single day. In addition, the assay may readily be scaled up to handle large numbers through automation.

A type of marker that better fits the needs of plant variety protection for high-quality, highly reproducible genetic information is the simple sequence repeat (SSR). SSRs, also known as microsatellites, are tandem repeats of short (2–6 bp) DNA sequences (Litt and Luty, 1989) that are abundant and uniformly dispersed in plant genomes (Lagercrantz et al., 1993; Wang et al., 1994). These repeats are highly polymorphic, even among closely related cultivars, due to a high frequency of mutations in the number of repeating units. Different SSR alleles may be detected at a locus by PCR using conserved DNA sequences flanking the SSR as primers. The sequences of pairs of primers flanking an SSR locus constitute a sequence-tagged site that can be documented and disseminated, thus allowing independent analysis of the same genetic locus by researchers anywhere in the world. SSRs are codominant and simple Mendelian segregation occurs (Saghai-Maroof et al., 1994). Morell et al. (1995) state that "STS markers, such as microsatellites, provide robust markers that may prove useful as primary characters for Plant Variety Registration."

The scoring of SSR alleles has been greatly simplified and improved by the advent of semiautomated fluorescent detection and sizing of PCR products, utilizing the Perkin Elmer/Applied Biosystems GeneScan™[1] system. This technology utilizes fluorescent labeling of one of the pair of primers used to amplify a SSR locus. The resultant PCR product is then electrophoresed in a denaturing acrylamide gel. The fluorescent marker is detected as a fragment crossing the path of a laser that scans the gel. A set of size standards, labeled with a fluorescent dye of a different color, is incorporated into every lane of the gel, allowing for extremely precise determinations of fragment sizes. Allele sizes for each "DNA sample × primer pair" combination is then automatically collated and analyzed.

[1] Registered trademark of Perkin Elmer/Applied Biosystems, Inc., Foster City, CA.

An obstacle to widespread use of SSRs for multiple applications has been the cost of discovering and characterizing SSRs in the genome of each plant species of interest. This task generally requires the screening of genomic libraries and extensive sequencing, as well as the design, synthesis, and optimization of PCR primers. Recent results (Kresovich et al., 1995) indicated that a set of SSR-specific primers may be used among closely related species within a genus. This broader applicability is particularly relevant to the turfgrass industry, where multiple species and subspecies are commonly released for commercial use.

Additional cost savings may be realized in SSR-based genetic analysis by single-tube multiplexing. Recent work in human (Kimpton et al., 1993) and plant systems (Mitchell et al., 1997) has demonstrated that data from many different SSR markers can be obtained simultaneously from a single PCR reaction. For example, no interference among canola (*Brassica napus* L.) SSRs has been observed with up to 11 in a single reaction, as long as the design of primers utilizes similar annealing temperatures and avoids sequence similarities which may preclude desired primer binding. The ABI GeneScan™ system allows PCR primers to be labeled with three different fluorescent dyes, and products can be accurately sized in the range of 75–500 bp. However, because each primer produces PCR products over a range of sizes, care must be taken to label and group primers so that no overlap occurs between products from different primers labeled with the same dye.

An initial set of SSR markers has been developed for seashore paspalum (Liu et al., 1995). The objectives of this effort were to establish the discriminatory ability of the markers and to evaluate their collective use for cultivar identification and diversity assessment among a test array of elite seashore paspalum accessions.

METHODS

Plant Materials

Seashore paspalum ecotypes and cultivars were obtained from the collection maintained by R.R. Duncan, Department of Crop and Soil Sciences, University of Georgia, Georgia Experiment Station, Griffin, Georgia. Entry information is presented in Table 4.1.

DNA Extraction

DNA was extracted via a modified CTAB procedure (Saghai-Maroof et al., 1984; Colosi and Schaal, 1993). Leaf tissue harvested from greenhouse-grown plants was lyophilized, then ground in a mortar and pestle with liquid N_2. Powdered tissue (approximately 100 mg) was mixed with 300 μL CTAB

Table 4.1. Seashore Paspalum Test Array Selected for SSR Analysis.

Entry	Source
AP-10	Alden Pines Country Club, Bookeelia, FL Pine Island, N. Fort Meyers, FL[a]
AP-14	Alden Pines Country Club, Bookeelia, FL Pine Island, N. Fort Meyers, FL[a]
PI 509018-1	Argentina
Mauna Key	Hawaii, Big Island
Tropic Shore	Hawaii, Molokai and Oahu
Utah-1	Selection from Adalayd
Adalayd	Australia
Excalibur	Selection from Adalayd
Fidalayel	Selection from Adalayd
FR-1	Fairbanks Ranch Country Club, N. San Diego, CA

[a] Selection from Adalayd.

extraction buffer [1.5% CTAB, 75 m*M* Tris (pH 8.0), 100 m*M* EDTA, 1.05 *M* NaCl, 0.75% PVP (40K)], incubated at 65°C for 15 min, then extracted twice with chloroform/isoamyl alcohol (24:1). DNA was precipitated by addition of 300 μL CTAB precipitation buffer [1% CTAB, 50 m*M* Tris (pH 8.0), 10 m*M* EDTA] and 600 μL isopropanol, incubated at room temperature for 30 min, then centrifuged at 12,000 rpm for 10 min. DNA pellets were resuspended in TE buffer, incubated with RNase A (1 μg/mL) for 30 min at 37°C, and extracted with phenol/chloroform/isoamyl alcohol (25:24:1). DNA was precipitated at –20°C in 70% ethanol, dried, and resuspended in HPLC-grade water at a concentration of 10 ng/μL.

SSR Primer Design and Preparation

Seashore paspalum SSRs were characterized previously by Liu et al. (1995). Primer sequences, the SSRs, annealing temperatures, and fluorescent labels are shown in Table 4.2. Fluorescent oligonucleotides were provided by Perkin Elmer, Foster City, California. Primers were labeled at the 5' end by incorporation of a phosphoramidite tagged with a fluorescent dye, either 6-carboxyfluorescein (6-FAM), tetrachloro-6-carboxyfluorescein (TET), or hexachloro-6-carboxyfluorescein (HEX), during DNA synthesis. Only one of the oligonucleotides (the “forward” primer) from each primer set was labeled.

DNA Amplification

PCR reactions were performed in 25 μL volumes containing 25 ng of template DNA, 1X PCR Buffer II (Perkin Elmer), 0.25 m*M* dNTPs, 1.5 m*M* $MgCl_2$,

Table 4.2. Characteristics of Dye-Labeled SSR Primers for *Paspalum Vaginatum*.[a]

SSR Primer	Label	Sequence	T_m[b]	Repeat	Allele Size Range (bp)[c]
Pv-3	Hex	F: TATGGACCGACTGCATGATTCTT			
		R: GTAGCTAGGTGAGAGGCATTC	48	$(AC)_{14}$	145–235
Pv-11	Tet	F: AGGTTTGTAGGTTGGGTGCAACTGA			
		R:TTGGGCGGCGGAGGGTAAT	60	$(AG)_{13}$	75–140
Pv-35	Hex	F: TCGAAATCGAAAAAGAAGATCGTTC			
		R: GGCGCCAGCTACAAGGTTAG	54	$(AG)_{21}$	95–130
Pv-51	Tet	F: TCCCATCATCAGTTCTTCCAATC			
		R: GCCCTGTGCTATTATTCATCATCTT	55	$(AG)_{13}$	90–130
Pv-53	Fam	F: CTCGGAAACCGCAGCTCA			
		R: GCTCCGCCTCCTCTATTCCA	55	$(AG)_{15}$	95–135

[a] Data of Liu et al. (1995).

[b] Annealing temperature (T_m) was calculated by the nearest-neighbor method using Right Primer™* from BioDisk Software.

[c] Size in bp was computed using the "local Southern" algorithm.

* Registered Trademark of BioDisk Software, Inc., San Francisco, California.

1.25 U T*aq* polymerase (AmpliTaq, Perkin Elmer), and 10 p*M* of each primer pair. Temperature cycling was done using the GeneAmp 9600 (Perkin Elmer) with 1 sec ramp times. The amplification profile consisted of initial denaturation of the template DNA at 95°C for 4 min, 25 cycles of 95°C for 1 min, 55°C for 2 min, and 72°C for 2 min. In the final PCR cycle, the extension time at 72°C was increased to 10 min (modified from Ziegle et al., 1992).

Electrophoresis and Detection

Samples containing 0.5 µL of the PCR products, 0.5 µL GENESCAN 500 internal lane standard labeled with N,N,N′,N′-tetramethyl-6-carboxyrhodamine (TAMARA) (Perkin Elmer) and 50% formamide were heated at 92°C for 2 min, placed on ice, then loaded on 6% denaturing acrylamide gels (24 cm well-to-read format). DNA samples were electrophoresed (29 watts) for 7 hr on an ABI model 373A automatic DNA sequencer/fragment analyzer equipped with GENESCAN 672 software v. 1.2 (Perkin Elmer). DNA fragments were sized automatically using the "local Southern" sizing algorithm.

Data Analysis

The level of polymorphism of each marker was estimated with a diversity index D [$D=1-\sum p_i^2$, in which p_i is the frequency of the ith SSR allele (Saghai-Maroof et al., 1994)]. Average similarity among accessions for each marker was calculated simply as 1-D. Pairwise distances were calculated both by a simple pairwise band sharing index and by the "delta mu" program developed for analysis of microsatellite data by Goldstein et al. (1995).

RESULTS

DNA profiles were generated for each of the ten seashore accessions (Table 4.3) based on summarization of data from 50 electropherograms (10 accessions × five SSRs). The number of fragments detected (average of 8.4) and the diversity index (average of 0.79) for each primer pair were quite high, similar to SSRs in soybean (*Glycine max* L.) (Rongwen et al., 1995) but less than that detected for SSRs in barley (*Hordeum vulgare* L.) (Saghai-Maroof et al., 1994). Surprisingly, three cultivars — Excalibur, Fidalayel, and FR-1 — had identical DNA profiles based on the five markers. Excluding Fidalayel and FR-1 (DNA profile represented by Excalibur), pairwise comparisons showed that the five SSR markers provided enough resolution to discriminate among — and partition variation for — the seashore paspalum accessions. Overall fragment sharing for the SSRs varied from 0.16–0.26 (Table 4.3). Figure 4.1 shows a dendrogram that depicts the genetic relationship among the ten cultivars/

Table 4.3. SSR Fragment Size Data for Ten Seashore Paspalum Cultivars.

	SSR Primers				
Entry	**Pv-3**	**Pv-11**	**Pv-35**	**Pv-51**	**Pv-53**
Excalibur	174	83	121	111	110
	184	105	133	121	
Fidalayel	174	83	121	111	110
	184	105	133	121	
FR-1	174	83	121	111	110
	184	103	133	121	
		105			
Mauna Key	184	83	109	113	108
		97	123		116
Tropic Shore	168	75	119	111	104
	180	83	123	113	116
	200	95	129		
		105			
Utah-1	170	83	121	115	106
		99	123	121	
		103			
Adalayd	174	90	121	111	110
	184	105	133	121	
PI 509018-1	174	85	115	108	112
	200	93	127	119	116
		99		121	
AP-10	184	87	127	111	112
	200	99	135	121	114
		105			
AP-14	174	89	107	119	112
	200	97	121	121	116
		101			
Number of fragments	6	13	10	6	7
Diversity index	0.76	0.87	0.84	0.74	0.76
Average similarity	0.24	0.13	0.16	0.26	0.24

ecotypes. Excalibur/FR-1/Fidalayel were genetically identical and obviously selections from Adalayd. The three ecotypes currently being evaluated on golf courses (AP10, AP14, PI509018-1) were genetically distinct, but distantly related. Utah, Mauna Key, and Tropic Shore were unique accessions that clustered together and were not closely related to the other seven accessions. Among the seashore paspalum accessions, the greatest similarities among entries were 85% for Adalayd and Excalibur, 55% for PI 509018-1 and AP-14, and 45% for PI 509018-1 and AP-10 (Table 4.4).

For breeding purposes, seashore paspalum is hypothesized to be a diploid; however, in some accessions more than two fragments were detected among

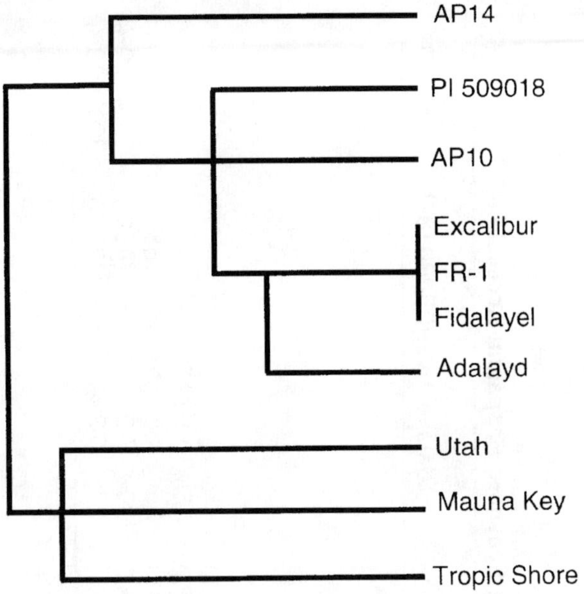

Figure 4.1. Dendrogram exhibiting the genetic relationship among ten paspalum accessions.

four of the five SSRs investigated (Table 4.3). Possible explanations for this phenomenon include a more complex genetic system than hypothesized (ploidy level), genetic mosaicism, contamination of greenhouse pots with multiple accessions, and amplification of unrelated genomic sites by the SSR primers or other PCR artifacts. Further investigation to resolve this key point is clearly warranted.

DISCUSSION

DNA typing of seashore paspalum ecotypes and cultivars using SSR markers and fluorescent semiautomated detection can potentially provide an accurate, reproducible, quick, low-cost method for cultivar identification and plant variety protection. In addition to solving the pragmatic problems associated with cultivar identification, genetic relationships between ecotypes and cultivars that may interest and be of value to breeders and the turfgrass industry can be revealed. As noted previously, seashore paspalum is thought to behave as a diploid, yet in some accessions more than two fragments were produced by four of the five SSRs. These ambiguous data underscore the fact that the basic biology and genetics of *P. vaginatum* require more study.

A most surprising result encountered in this study was that three seashore paspalum accessions, Excalibur, Fidalayel, and FR-1, were genetically identi-

Table 4.4. Pairwise SSR Fragment Sharing Similarity (Difference) Among Seashore Paspalum Accessions.

	Excalibur	Mauna Key	Tropic Shore	Utah-1	Adalayd	PI 509018-1	AP-10
Excalibur							
Mauna Key	15% (85%)						
Tropic Shore	20% (80%)	30% (70%)					
Utah-1	30% (70%)	15% (85%)	10% (90%)				
Adalayd	85% (15%)	10% (90%)	15% (85%)	20% (80%)			
PI 509018-1	15% (85%)	10% (90%)	20% (80%)	15% (85%)	15% (85%)		
AP-10	35% (65%)	10% (90%)	20% (80%)	15% (85%)	35% (65%)	45% (55%)	
AP-14	30% (70%)	15% (85%)	15% (85%)	20% (80%)	30% (70%)	55% (45%)	30% (70%)
Average	33% (67%)	13% (87%)	16% (84%)	18% (82%)	27% (73%)	50% (50%)	30% (70%)

cal. Estimates of the number of SSR markers required for cultivar identification are based on the combined discrimination power provided by multiple markers. This discrimination ability is dependent on the assumption that alleles of one marker are not linked to alleles of other markers across any significant subfractions of a collection (linkage disequilibrium). Provided that no linkage disequilibrium occurs, then the discrimination power of the assay increases exponentially with each additional marker. For example, if each marker has a diversity of 0.5 (which corresponds to a 50% chance of discriminating two unrelated plants), then the total discrimination power can be calculated with the formula $X = (1/(1\text{-}D))^n$, where X is the number of unique genotypes that can be generated (equivalent to discrimination power), D is the average diversity of the markers, and n is the number of marker loci assayed. Therefore, a single marker of diversity 0.5 will provide $(1/0.5)^1 = 2$ genotypes, 10 markers will yield $(2)^{10} = 1024$ genotypes, and 20 markers will yield $(2)^{20} = > 1$ million genotypes. The five seashore paspalum SSRs utilized in this study have an average diversity of 0.79, which provides a discrimination power of $(1/(1\text{-}0.79))^5 = 2448$, so the likelihood that the three cultivars, "Excalibur," "Fidalayel," and FR-1 all share the same DNA profile by chance is less than 1 in 500,000.

Two seashore paspalum ecotypes, AP-10 and AP-14, currently being evaluated on greens under golf course conditions, were found to be genetically distinct from each other (30% similarity) and from all other accessions examined (maximum similarity of 55% between AP-10 and PI 509018).

Though conducted in a rigorous manner, the SSR data generated for seashore paspalum accessions were not collected under conditions necessary for forensic applications. Nonetheless, with increased quality control, forensic standard data can be collected using existing technology. The further characterization of each SSR marker must be established, particularly in reference to the range of fragment sizes, their frequencies in wild and cultivated populations, and their genetic basis. A more thorough study is currently under way to provide this greatly needed information. Nonetheless, this preliminary study suggests that SSR markers ultimately will provide readily useful and discriminatory tools for turfgrass curators, breeders, and the user community.

REFERENCES

Brown, S.M., A.K. Szewc-McFadden, and S. Kresovich. 1995. Development and application of simple sequence repeat (SSR) loci for plant genome analysis. In *Methods of Genome Analysis in Plants*, Jauhar, P., Ed., CRC Press, Boca Raton, FL.

Caetano-Anollés, G. 1994. MAAP — A versatile and universal tool for genome analysis. *Plant Molecular Biology,* 25:1011–1026.

Caetano-Anollés, G., L.M. Callahan, P.E. Williams, K.R. Weaver, and P.M. Gresshoff. 1995. DNA amplification fingerprinting analysis of bermudagrass

(*Cynodon*): genetic relationships between species and interspecific crosses. *Theoretical Applied Genetics,* 91:228–235.

Colosi, J.C. and B.A. Schaal. 1993. Tissue grinding with ball bearings and vortex mixer for DNA extraction. *Nucleic Acids Research,* 21:1051–1052.

Goldstein, D.B., A. Ruíz-Linares, M. Feldman, and L.L. Cavalli-Sforza. 1995. An evaluation of genetic distances for use with microsatellite loci. *Genetics,* 139:463–471.

Graham, G.C., R.J. Henry, and R.J. Redden. 1994. Identification of navy bean varieties using random amplification of polymorphic DNA. *Australian J. Experimental Agriculture* 34:1173–1176.

Huff, D.R., R. Peakall, and P.E. Smouse. 1993. RAPD variation within and among natural populations of outcrossing buffalograss [*Buchloe dactyloides* (Nutt) Engelm]. *Theoretical Applied Genetics,* 86:927–934.

Kimpton, C.P., P. Gill, A. Walton, A. Urquhart, E.S. Millican, and M. Adams. 1993. Automated DNA profiling employing multiplex amplification of short tandem repeat loci. *PCR Methods Application,* 3:13–22.

Kresovich, S., A.K. Szewc-McFadden, S.M. Bliek, and J.R. McFerson. 1995. Abundance and characterization of simple sequence repeats (SSRs) isolated from a size-fractionated genomic library of *Brassica napus* L. (rapeseed). *Theoretical Applied Genetics,* 91:206–211.

Lagercrantz, U., H. Ellegren, and L. Andersson. 1993. The abundance of various polymorphic microsatellite motifs differs between plants and vertebrates. *Nucleic Acids Research,* 21:1111–1115.

Litt, M. and J.A. Luty. 1989. A hypervariable microsatellite revealed by *in vitro* amplification of a dinucleotide repeat within the cardiac muscle actin gene. *American Journal Human Genetics,* 44:397–401.

Liu, Z.W., R.L. Jarret, R.R. Duncan, and S. Kresovich. 1994. Genetic relationships and variation of ecotypes of seashore paspalum (*Paspalum vaginatum* Swartz) determined by random amplified polymorphic DNA (RAPD) markers. *Genome,* 37:1011–1017.

Liu, Z.W., R.L. Jarret, S. Kresovich, and R.R. Duncan. 1995. Characterization and analysis of simple sequence repeat (SSR) loci in seashore paspalum (*Paspalum vaginatum* Swartz). *Theoretical Applied Genetics,* 91:47–52.

Mitchell, S.E., S. Kresovich, C.A. Jester, C.J. Hernandez, and A.K. Szewc-McFadden. 1997. Application of multiplex PCR and fluorescence-based, semi-automated sizing technology for genotyping plant genetic resources. *Crop Sci.* 37:

Morell, M.K., R. Peakall, R. Appels, L.R. Preston, and H.L. Lloyd. 1995. DNA profiling techniques for plant variety protection. *Australian Journal Experimental Agriculture,* 35:807–819.

Novy, R.G., C. Kobak, J.C. Goffreda, and N. Vorsa. 1994. RAPDs identify varietal misclassification and regional divergence in cranberry [*Vaccinium macrocarpon* (Ait.) Pursh]. *Theoretical Applied Genetics,* 88:1004–1010.

Rongwen, J., M.S. Akkaya, A.A. Bhagwat, U. Lavi, and P.B. Cregan. 1995. The use of microsatellite DNA markers for soybean genotype identification. *Theoretical Applied Genetics,* 90:43–48.

Saghai-Maroof, M.A., R.M. Biyashev, G.P. Yang, Q. Zhang, and R.W. Allard. 1994. Extraordinarily polymorphic microsatellite DNA in barley: species diversity, chromosomal locations, and population dynamics. *Proceedings National Academy of Sciences USA,* 91:5466–5470.

Saghai-Maroof, M.A., K.M. Soliman, R.A. Jorgensen, and R.W. Allard. 1984. Ribosomal DNA spacer length polymorphisms in barley: Mendelian inheritance, chromosomal location, and population dynamics. *Proceedings National Academy of Sciences USA,* 81:8014–8018.

Siedler, H., M.M. Messmer, G.M. Schachermayr, H. Winzeler, M. Winzeler, and B. Keller. 1994. Genetic diversity in European wheat and spelt breeding material based on RFLP data. *Theoretical Applied Genetics,* 88:994–1003.

Smith, J.S.C. and O.S. Smith. 1992. Fingerprinting crop varieties. *Advances in Agronomy,* 47:85–140.

Tancred, S.J., A.G. Zeppa, and G.C. Graham. 1994. The use of the PCR-RAPD technique in improving the plant variety rights description of a new Queensland apple (*Malus domestica*) cultivar. *Australian Journal Experimental Agriculture,* 34:665–667.

Thomas, M.R., P. Cain, and N.S. Scott. 1994. DNA typing of grapevines — A universal methodology and database describing cultivars and evaluating genetic relatedness. *Plant Molecular Biology,* 25:939–949.

Virk, P.S., H.J. Newbury, M.T. Jackson, and B.V. Ford-Lloyd. 1995. The identification of duplicate accessions within a rice germplasm collection using RAPD analysis. *Theoretical Applied Genetics,* 90:1049–1055.

Wang, Z., J.L. Weber, G. Zhong, and S.D. Tanksley. 1994. Survey of plant short tandem DNA repeats. *Theoretical Applied Genetics,* 88:1–6.

Weaver, K.R., L.M. Callahan, G. Caetano-Anollés, and P.M. Gresshoff. 1995. DNA amplification fingerprinting and hybridization analysis of centipedegrass. *Crop Science,* 35:881–885.

Wu, L. and H. Lin. 1994. Identifying buffalograss [*Buchloe dactyloides* (Nutt) Engelm] cultivar breeding lines using random amplified polymorphic DNA (RAPD) markers. *Journal of American Society Horticultural Science,* 119:126–130.

Yamagishi, M. 1995. Detection of section-specific random amplified polymorphic DNA (RAPD) markers in *Lilium. Theoretical Applied Genetics,* 91:830–835.

Yang, X. and C.F. Quiros. 1993. Identification and classification of celery cultivars with RAPD markers. *Theoretical Applied Genetics,* 86:205–212.

Ziegle, J.S., Y. Su, K.P. Corcoran, L. Nie, P.E. Mayrand, L.B. Hoff, L.J. McBride, M.N. Kronick, and S.R. Diehl. 1992. Application of automated DNA sizing technology for genotyping microsatellite loci. *Genomics,* 14:1026–1031.

Part 2

Biological Control, Including Endophyte Strategies

Chapter 5

Microbial Mechanisms of Biological Disease Control

E.B. Nelson

BIOLOGICAL CONTROL IN TURFGRASS SYSTEMS

Traditional turfgrass management programs have relied heavily on fungicide applications for disease control. It has only been in recent years that a more visible trend toward nonchemical alternatives has become apparent. Not only are turfgrass managers demanding alternatives, but an increasing number of research laboratories around the world are now focusing efforts on biological methods of disease control. In a little over ten years, biological control of turfgrass diseases has come from being simply a scientific curiosity to an accepted and viable technology.

The more common approaches for implementing biological control strategies have involved the use of either microbial inoculants or organic amendments (Nelson, 1996; Nelson et al., 1994). A key element of both strategies is maintaining elevated populations and activity of microbes in treated soils. Since the use of organic amendments for the biological control of turfgrass diseases has been reviewed recently (Nelson, 1996), it will not be covered here. Instead, this review will focus on the use of microbial inoculants for turfgrass disease control and some of the applications of molecular and cellular biology for understanding biological control processes in these systems.

MICROBIAL INOCULANTS FOR TURFGRASS DISEASE CONTROL

Microbial inoculants have been used in turfgrasses for such things as thatch reduction (Mancino et al., 1993), fertilizer enhancement (Peacock and Daniel,

1992), insect control (Villani, 1995), weed control (Zhou and Neal, 1995), and, perhaps most importantly, disease control (Table 5.1). Over the years, numerous microbial disease control agents have been described. The most commonly studied microorganisms for biological disease control in turfgrasses have been species of the bacterial genera *Pseudomonas, Enterobacter,* and *Streptomyces,* as well as species of the fungal genera *Trichoderma, Typhula,* and *Gliocladium.* The majority of studies with these inoculants have been largely descriptive, documenting control efficacy and, in some cases, emphasizing ecological relationships. Few studies have focussed on biological control mechanisms.

One of the more thorough studies with microbial inoculants has been on the biological control of *Typhula* blight with strains of *Typhula phacorrhiza. T. phacorrhiza* is closely related to the fungi that cause *Typhula* blight, but it is not pathogenic to turfgrasses (Burpee et al., 1987) and specific isolates are highly disease-suppressive (Nelson et al., 1994). Inverse relationships between the concentration of inoculum of *T. phacorrhiza* applied to turf in November and the intensity of *Typhula* blight the following March have been observed (Nelson et al., 1994). Application of *T. phacorrhiza* at a rate of 7.0×10^3 colony-forming units (cfu)/m^2 of turf provided control of *Typhula* blight equal to that achieved with the fungicide pentachloronitrobenzene (PCNB) applied at 30 kg a.i./ha. *T. phacorrhiza* also survives and reproduces in turfgrass thatch and soil, and suppresses *Typhula* blight up to 16 months after two annual applications of infested grain (Lawton and Burpee, 1990).

Another well-studied microbial inoculant for turfgrasses is *Trichoderma harzianum* strain 1295-22 (Harman and Lo, 1996; Lo et al., 1996, 1997). This biocontrol fungus is an effective control agent for dollar spot, brown patch, and *Pythium* root rot diseases on creeping bentgrass. One of the more intriguing properties of this organism is its ability to persist in the rhizosphere of creeping bentgrass. Monthly applications of *T. harzianum* have been effective in maintaining populations at levels of nearly 10^6 cfu/g of thatch/soil. In some experiments, populations increased with each successive application (Lo, Nelson, and Harman, unpublished). *T. harzianum* was also shown to overwinter at population levels between 10^5–10^6 cfu/g — levels adequate to achieve a certain level of biological control. However, if populations fall below 10^5 cfu/g, biological control efficacy is lost (Lo et al., 1997). This strain is now commercially available as BioTrek 22G™,[1] the first biological disease control agent for turfgrasses registered with the U.S. EPA (Harman and Lo, 1996).

MECHANISMS OF BIOLOGICAL CONTROL

Over the past few years, there has been much interest in understanding the mechanisms by which microbial inoculants suppress pathogens and their re-

[1] BioTrek 22G is a registered trademark of Wilbur-Ellis, Fresno, California.

Table 5.1. Microbial Inoculants Studied for Turfgrass Disease Control.

Biocontrol Organism	Target Diseases	Lab (L)/Field (F) Study	References
Fungal Agents			
Acremonium spp.	Dollar spot	F	(Goodman and Burpee, 1991)
Fusarium heterosporum	Dollar spot	L	(Goodman and Burpee, 1991)
Gaeumannomyces spp.	Take-all patch	F	(Wong and Worrand, 1989)
Gliocladium virens	Brown patch	L/F	(Giesler et al., 1993; Haygood and Mazur, 1990; Haygood and Walter, 1991; Yuen et al., 1994)
Laetisaria spp.	Brown patch	L/F	(Sutker and Lucas, 1987)
Phialophora radicicola	Take-all patch	F	(Deacon, 1973; Wong and Siviour, 1979)
Rhizoctonia spp.	Brown patch	L/F	(Burpee and Goulty, 1984; Yuen and Craig, 1992; Yuen et al., 1994)
Trichoderma spp.	Southern blight	F	(Punja et al., 1982)
	Brown patch	L	(O'Leary et al., 1988)
	Typhula blight	L	(Harder and Troll, 1973)
T. hamatum	*Pythium* blight	F	(Rasmussen-Dykes and Brown, 1982)
	Dollar spot	F	(Grebus et al., 1995; Grebus et al., 1994; Segall, 1995)
T. harzianum	Dollar spot	L/F	(Lo et al., 1996; Lo et al., 1997)
	Brown patch	L/F	(Lo et al., 1996; Lo et al., 1997)
	Pythium root rot	L/F	(Lo et al., 1996; Lo et al., 1997)
Typhula phacorrhiza	*Typhula* blight	L/F	(Burpee, 1994; Burpee et al., 1987; Lawton and Burpee, 1990; Lawton et al., 1987)
Bacterial Agents			
Enterobacter cloacae	Dollar spot	F	(Nelson and Craft, 1991)
	Pythium blight	L	(Nelson and Craft, 1992)

Table 5.1. Microbial Inoculants Studied for Turfgrass Disease Control (Continued).

Biocontrol Organism	Target Diseases	Lab (L)/Field (F) Study	References
Bacterial Agents (Continued)			
Flavobacterium balustinum	Dollar spot	F	(Grebus et al., 1995; Grebus et al., 1994; Segall, 1995)
Pseudomonas spp.	Take-all patch	F	(Lucas and Sarniguet, 1991; Lucas et al., 1992a,b; Lucas et al., 1991; Sarniguet and Lucas, 1991)
	Pythium blight	F	(Wilkinson and Avenius, 1985)
P. fluorescens	Take-all patch	F	(Baldwin et al., 1991)
	Dollar spot	L	(Hodges et al., 1994)
	Leaf spot	L	(Hodges et al., 1994)
P. lindbergii	Dollar spot	L	(Hodges et al., 1994)
	Leaf spot	L	(Hodges et al., 1994)
P. putida	Take-all patch	L/F	(Wong and Baker, 1984; Wong and Baker, 1985)
Serratia marcescens	Summer patch	L/F	(Kobayashi and El-Barrad, 1996)
Xanthomonas maltophilia	Summer patch	L/F	(Kobayashi et al., 1995)
Various bacteria	Summer patch	L/F	(Thompson and Clarke, 1992; Thompson et al., 1993)
Actinomycete Agents			
Streptomyces spp.	Leaf spot	L	(Hodges et al., 1993)
	Dollar spot	F	(Reuter et al., 1991; Schumann and Reuter, 1993)
	Brown patch	F	(Reuter et al., 1991)
Various species	*Pythium* root rot	L/F	(Stockwell et al., 1994)
Unreported Identity	*Pythium* blight	F	(Nelson and Craft, 1992; Sanders and Soika, 1991; Soika and Sanders, 1991; Soika and Sanders, 1992)

spective diseases. One of the more important practical reasons for such studies is the development of approaches for predicting the behavior of microbial inoculants. Due to the extremely close link between microbial function and performance, one cannot readily predict the behavior of introduced biological control agents without an understanding of the mechanisms involved in pathogen or disease suppression. It also follows that the performance of biological control agents can be enhanced if their physiology and ecology is more clearly understood.

In studies of biological control mechanisms it has become apparent that biocontrol activity is not a simple process mediated by a single microbial metabolite. The capacity of an organism to simply produce a fungicidal compound or another biologically active compound in and of itself is not always sufficient to explain biological control activity. Rather, biological control activity involves a series of events not unlike those involved in plant pathogenesis or in the establishment of symbiotic plant-microbe associations. Biological control mechanisms can be thought of as a series of traits expressed synchronously or in a controlled sequence. Therefore, as a first step in understanding biocontrol processes, it is important to assemble the traits organisms must possess to function as effective biological control agents. While many, if not most, biological control microbes may suppress diseases through a variety of different mechanisms (Ownley et al., 1992), a few key traits appear to be common to biological control mechanisms but are likely to differ somewhat among various microbes.

MICROBIAL TRAITS IMPORTANT TO BIOLOGICAL CONTROL

Essentially all of the work on microbial traits related to biological control processes has come from studies with crop plants other than turfgrasses. However, many of these traits and processes may also play important roles in turfgrass ecosystems. Of the traits common to soil microbes, there are at least five that have been consistently linked with biological control activity. These include (1) the production of toxic metabolites such as antibiotics and inhibitory volatiles; (2) microbial resource competition; (3) hyperparasitism; (4) induction of systemic plant resistance; and (5) rhizosphere competence. Not all of these traits are mutually exclusive, and some biological control organisms may require various combinations of these traits for biological control activity to be expressed.

Production of Inhibitory Metabolites

The most commonly studied trait related to biological control activity, particularly in bacterial systems, has been antibiotic biosynthesis. Antibiotic substances produced in particular by species of *Pseudomonas* are known to play key roles in the biological control of *Pythium* and *Rhizoctonia* diseases. The more commonly described antibiotics involved in these biological control processes include pyrrolnitrin (Burkhead et al., 1994; Chernin et al., 1996; Hill et al., 1994; Homma, 1984; Homma and Suzui, 1989; Sarniguet and Loper, 1994), pyoluteorin (Howell and Stipanovic, 1980; Kraus and Loper, 1995; Maurhofer et al., 1994), 2,4-diacetylphloroglucinol (Carroll et al., 1995; Fenton et al., 1992; Keel, 1990; Keel et al., 1992; Nowak-Thompson et al., 1994), phenazine-1-carboxylic acid (Georgakopoulos et al., 1994a,b; Mazzola et al., 1992; Pierson et al., 1994; Pierson and Thomashow, 1991; Slininger and Jackson, 1992; Slininger and Sheawilbur, 1995; Thomashow, 1990; Thomashow and Pierson, 1991; Thomashow and Weller, 1988), and oomycin A (Gutterson et al., 1988; Gutterson et al., 1986; Howie and Suslow, 1986; Howie and Suslow, 1991; James and Gutterson, 1986; Tucker et al., 1988). Other antibiotics have been described that may be important in other pathosystems.

The majority of studies on antibiotic biosynthesis and its relation to biological control have been limited primarily to fluorescent *Pseudomonas* species, with the majority of research encompassing only 5 to 10 strains of *Pseudomonas fluorescens*, a few strains of *P. aureofaciens*, and a couple of strains of *P. putida*. Even though many studies have been limited to these few strains of fluorescent pseudomonads, important concepts have been generated, particularly with regard to the regulation of antibiotic biosynthesis (Laville et al., 1992; Sarniguet and Loper, 1994). Despite this, a broader diversity of biocontrol agents must be studied against a wider array of pathogens before a model of antibiotic biosynthesis and regulation can be developed and the role of this trait in biological control processes can be understood.

Fungal biological control agents, particularly in the genera *Trichoderma* and *Gliocladium*, synthesize an endless list of antifungal antibiotics with important roles in biological control processes. Traditionally the role of fungal antibiotics in biological control processes has been difficult to assess. Much of our information relative to fungal antibiosis and biological control has come from the isolation, identification, and bioassay of metabolites released into culture media (Claydon, 1987; Howell and Stipanovic, 1983; Wilhite et al., 1994). Recently Wilhite et al. (1994) were able to develop gliotoxin mutants of the biocontrol fungus, *Gliocladium virens*, and show that gliotoxin production was involved in the biological control of *Pythium* and *Rhizoctonia* diseases. This is the strongest evidence to date for the role of antibiotics in biological control by *Gliocladium*.

With both fungal and bacterial biological control organisms, volatile microbial inhibitors also have been implicated in biological control processes. Species of *Pseudomonas* have been shown to produce hydrogen cyanide under some conditions (Ahl et al., 1986; Ross and Ryder, 1994; Voisard et al., 1989) and this has been linked to biological control activity. In *in vitro* tests, *Enterobacter cloacae* was shown to produce ammonia suppressive to *Pythium ultimum* and *Rhizoctonia solani* (Howell et al., 1988).

Microbial Competition

Microbial competition has often been proposed as a mechanism of biological control, yet no definitive evidence to support this contention currently exists. Although intuitively, competition should be an important mechanism of biological control, the complexities of the soil environment make the proof of competition mechanisms difficult to obtain. There is now some empirical evidence that competition for iron and pathogen germination stimulants might play a role in biological control.

Siderophores

Siderophores are low-molecular-weight iron chelates that are produced by many soil microbes under iron-limiting conditions. Siderophores chelate ferric iron and serve as a major vehicle for iron transport into microbial cells. Nearly all organisms produce siderophores, but those produced by species of *Pseudomonas* and enteric bacteria generally have higher affinities for iron than do other fungal siderophores. Siderophores have been studied in relation to biological control since the initial report back in 1980 (Kloepper and Leong, 1980). Good reviews of the subject are available (Buyer et al., 1994; Leong, 1986; Loper and Buyer, 1991).

In a number of specific pathosystems, the biological control of soilborne pathogens has been attributed to the production of siderophores (Becker and Cook, 1988; Buysens et al., 1996; Elad and Barak, 1983; Kloepper and Leong, 1980; Lemanceau et al., 1992; Loper, 1988), whereas in other cases no role for siderophores can be found (Ahl et al., 1986; Kraus and Loper, 1992; Paulitz and Loper, 1991; Trutmann and Nelson, 1992). Although siderophore competition has generally been considered a direct form of biological control, it is possible that some siderophores may be acting indirectly by enhancing natural plant defense mechanisms under iron-limiting conditions (Buysens et al., 1996; Leeman et al., 1996). More work is needed to resolve the functional relationships between siderophore production and biological control.

Inactivation of Plant Exudates

Seed and root exudates play an important role in the initiation of soilborne plant diseases by serving as stimulants of fungal propagules (Curl and Truelove, 1986; Nelson, 1990). Without the release of stimulatory molecules in these exudates, pathogenic relationships between plants and soil pathogens do not occur. Recently, there has been interest in understanding the interaction of biocontrol organisms with stimulatory components of seed and root exudates. Since soilborne fungal pathogens are highly dependent on exudate molecules to initiate plant infections, microbial interference with the production and activity of exudate stimulants could be an effective mechanism of biological control among seed-applied spermosphere bacteria.

A limited amount of experimental evidence supports such a mechanism of biological control (Ahmad and Baker, 1988; Fukui et al., 1994; Gorecki et al., 1985; Maloney et al., 1994; Nelson, 1990; Paulitz, 1991; Paulitz et al., 1992), with some of the studies including turfgrass systems (van Dijk, 1995). *Trichoderma* spp. have been shown to reduce the levels of certain stimulatory exudate components (Gorecki et al., 1985; Nelson, 1990; Paulitz, 1991; Paulitz et al., 1992) and to reduce responses of *Pythium ultimum* to plants, even in the apparent absence of any direct interaction with the pathogen (Ahmad and Baker, 1988).

The ability of bacterial biocontrol agents to reduce propagule germination of *Pythium* spp. in response to plants has been related to biocontrol efficacy. A significant correlation was observed between the ability of various bacterial strains to inhibit *Pythium* seed rot of cucumber and their ability to inhibit *P. aphanidermatum* oospore germination in the rhizosphere (Elad and Chet, 1987). Effective bacterial strains inhibited oospore germination by as much as 57% while ineffective strains inhibited oospore germination by only 13–20%. There was no direct interaction between bacteria and oospores, and there was no evidence for the bacterial production of inhibitory metabolites. Similar relationships have been observed in other pathosystems (Elad and Baker, 1985). Even stronger correlations between inactivation of exudate stimulants and *Pythium* biocontrol activity have been obtained with *Enterobacter cloacae* (Maloney et al., 1994; van Dijk, 1995) (described in detail below).

General Nutrient Competition

Few studies have effectively demonstrated a role for general nutrient competition in biological control processes. However, studies with *Typhula phacorrhiza* (Burpee et al., 1987; Lawton and Burpee, 1990; Lawton et al., 1987) indicate that some form of nutrient competition may be responsible for the observed biological control. Laboratory studies indicate that isolates of *T. phacorrhiza* do not produce antibiotics, and the fungus is not parasitic on other species of *Typhula*. However, a significant reduction in growth of *T. incarnata* and *T. ishikariensis* is observed on media previously exposed to *T. phacorrhiza* (Burpee et al., 1987). Applications of incremental increases in nutrients to the media, after exposure to *T. phacorrhiza*, resulted in significant linear increases in colony diameters of *T. incarnata* and in dry weights of sclerotia of *T. incarnata* and *T. ishikariensis*. These results suggest that, during saprophytic phases of growth, isolates of *T. phacorrhiza* may competitively utilize nutrients essential for the growth and development of pathogenic species of *Typhula*.

Mycoparasitism and the Production of Cell Wall–Degrading Enzymes

Mycoparasitism is a complex process by which biocontrol fungi may attack pathogenic fungi and involves the following steps, as inferred from *in vitro* studies (primarily from Chet, 1987): (1) the biocontrol fungi grow tropically toward the target fungi, (2) hyphae of the biocontrol fungi bind to lectins on the surface of the target fungi via attachment of carbohydrate receptors on the surface of the biocontrol fungus, (3) cell wall–degrading enzymes, such as chitinases, glucanases, and proteinases, are produced by the biocontrol fungus, destroying the cell integrity of the pathogenic fungus; these enzymes have recently been shown to be complex mixtures of synergistic proteins that act together against pathogenic fungi (Lorito et al., 1993), and (4) appressoria-like structures are produced that apparently initiate penetration of the target fungus by the biocontrol fungi. Although mycoparasitic structures have been observed on seeds of various crop plants treated with *Trichoderma harzianum* (Hubbard and Harman, 1983), to date, there is no proof that this process plays a significant role in the biological control of diseases by *Trichoderma* and other mycoparasitic fungi.

Definitive results concerning the role of mycoparasitism by *Trichoderma* in control of a range of pathogens should be available soon. The genes for cell wall–degrading enzymes are being isolated (Hayes et al., 1994), and the first chitinase deficient mutants have been prepared (Harman and Hayes, 1993). A gene encoding a proteinase enzyme has also been recently linked to mycoparasitic activity in *Trichoderma harzianum* (Geremia et al., 1993). Once a series of mutants deficient in specific cell wall–degrading enzymes have been prepared, as well as strains to which the genes have been restored, the role of the enzymes in biocontrol can be definitively assessed.

Bacteria may also be parasitic to pathogenic fungi, in which the biocontrol efficacy is mediated by the production of chitinase enzymes. *Serratia marcescens* is known not only for its chitinase production, but for its activity as a biological control agent for soilborne diseases, including summer patch on Kentucky bluegrass caused by *Magnaporthe poae* (Kobayashi and El-Barrad, 1996; Kobayashi et al., 1995). Much of the success of *S. marcescens* as a biological control agent is believed to result from its production of chitinolytic enzymes inhibitory to chitin-containing soilborne fungi. Additionally, species of *Enterobacter* also produce chitinolytic enzymes that have been directly associated with biological control activity (Chernin et al., 1995). A number of reports have been published in which chitinase genes from *S. marcescens* have been cloned and used for the development of transgenic bacteria and plants, each with enhanced suppression of soilborne diseases (Broglie et al., 1991; Cornelissen and Melchers, 1993; Fuchs et al., 1986; Haran et al., 1993; Howie et al., 1994; Jones et al., 1986; Koby et al., 1994; Shapira et al., 1989; Sundheim et al., 1988; Zhu et al., 1994).

It has long been the goal of many researchers in the field of biological control to develop more effective biocontrol strains or those that perform more consistently. The expression of chitinase genes in transgenic bacteria and fungi is currently the best example of the development of genetically engineered biological control agents for the suppression of soilborne diseases. This can serve as a suitable model for illustrating the possibilities of utilizing transgenic microorganisms for the control of turfgrass diseases.

Induced Systemic Resistance

There is now compelling evidence that a wide variety of microbial inoculants activate natural plant defense mechanisms capable of limiting infection and disease development at sites distant from the point of inoculation (Tuzun and Kloepper, 1994). This phenomenon, known as induced systemic resistance, is triggered when inoculated organisms come in contact with host plants and colonize roots. Since this is a relatively new area of biological control research, little is known about the actual mechanisms involved in resistance induction. However, it is clear that inoculation of plant roots with plant growth-promoting rhizobacteria will induce phytoalexin accumulation in plants along with a variety of chitinases, glucanases, and PR-proteins (Bol et al., 1996; Tuzun and Kloepper, 1994) typical of plant defense responses.

Current research in a number of laboratories around the world is focusing on the bacterial determinants responsible for triggering the resistance response. Recently, salicylic acid has been implicated as an important signal molecule capable of eliciting such responses (Bol et al., 1996). It is also known that a number of fluorescent *Pseudomonas* species produce salicylic acid (Buysens et al., 1996; Meyer et al., 1992). To date, only a couple of studies have demonstrated the effectiveness of induced systemic resistance under field conditions, and to my knowledge, all of this work has been conducted with dicotyledonous plants. Much remains to be learned about how this mechanism operates in turfgrass plants.

Rhizosphere Competence

Biological control efficacy is partially determined by the intrinsic ability of biocontrol organisms to grow on and colonize plant surfaces. This is particularly important for the biological control of root and crown diseases. The ability of microbes to establish and proliferate in the root zone is a trait referred to as rhizosphere competence (Ahmad and Baker, 1987a,b; Ahmad and Baker, 1988). The apparent rhizosphere competence of any microorganism is strongly affected by the method of application and by the physiological, ecological, and edaphic interactions that occur in the root zone habitat.

Rhizosphere competence is particularly important to microbial inoculants used for turfgrass disease control since turfgrasses generate a tremendous volume of roots, making root colonizing organisms particularly well-suited to a turfgrass habitat. Furthermore, the efficacy of biological control agents in turfgrass systems is highly dependent on their population level (Lo et al., 1996). In most biocontrol systems, populations of introduced inoculants must be at levels $>10^6$ cfu/g soil. The greater the level of rhizosphere competence, the more able the organism is to maintain high population levels.

Attempts have been made to improve rhizosphere competence in biological control fungi. *Trichoderma harzianum* strain 1295-22, the active organism in the product BioTrek 22G, is a product of protoplast fusion (Sivan and Harman, 1991). Combining parental strains with different biocontrol and root-colonizing properties, progeny with a high level of biological control activity as well as more complete root colonization were created.

MOLECULAR APPROACHES FOR UNDERSTANDING MICROBIAL MECHANISMS OF DISEASE CONTROL

With the refinement of molecular biological techniques and the adaptation of these techniques to biological control systems, many unanswered questions regarding the function of microbial inoculants and the role of specific microbial metabolites in complex biocontrol processes can now be addressed in a relatively straightforward manner. The utility of recombinant DNA techniques lies in the ability they give investigators to experimentally isolate DNA sequences that regulate biocontrol phenotypes and thus more definitively assess the role of these genes and their corresponding gene products in biocontrol processes. The basic strategy for determining the role of specific genetic elements in biocontrol processes is as follows: (1) development of an efficient assay to observe the desired biocontrol phenotype; (2) selection of wild-type strains with the desired biocontrol phenotype; (3) mutagenesis of strains; (4) screening of mutant strains for loss of the desired phenotype; (5) preparation of genomic or cDNA libraries from wild-type strains; and (6) complementation of phenotype-minus strains to restore the desired phenotype. This strategy is highly dependent on suitable and efficient gene transfer technologies. In bacterial systems, gene transfer occurs naturally through processes of conjugation, transformation and transduction. Since much of the basis for recombinant DNA technology has come from our understanding of bacterial genetics, it is not surprising that our knowledge of biocontrol processes mediated by bacteria is somewhat more advanced than that for fungal systems.

Transposon Mutagenesis

Perhaps the single most important discovery in bacterial genetics, making many of the applications of molecular biology possible, has been the discovery of transposable genetic elements, or transposons (Berg and Berg, 1987; Berg and Berg, 1983). Although transposons are widely distributed among prokaryotes, their presence in filamentous fungi is only beginning to be realized (Daboussi and Langin, 1994; Glayzer et al., 1995). The use of transposons in cell mutagenesis has been the most widely used strategy for generating random, unique, single-gene mutations in bacteria and has been used extensively in the study of molecular mechanisms of bacterial pathogenesis and biological control. The properties and applications of transposons have been reviewed at some length (Berg and Berg, 1987) and will not be covered in depth in this review.

The transposon Tn*5* has been the most widely used transposon for studies of biological control activity in gram-negative bacteria because of its wide host range and nearly random insertion into bacterial DNA. In mutagenizing cells for molecular genetic analyses, transposons carrying some sort of marker (usually antibiotic or heavy metal resistance) are delivered on plasmid vectors to host cells where they can transpose from the vector into the host cell genome. Detection of such transposition events depends on the host cell acquiring a new phenotype based on the selectable marker carried on the transposon. Many soilborne organisms carry multiple antibiotic resistance genes which may severely limit the useful choices of transposition vectors.

For transposons to be most useful, transposon delivery systems should be chosen that are biased for single transposition events and the subsequent elimination of the transpositional vector from the host cell. These events can be easily monitored. For example, transconjugants can be readily screened for resistance to the selectable marker, to ensure that the transposon has entered the host cell, and for sensitivity to the antibiotic marker carried on the vector, to ensure that the vector has not been retained in the transconjugant. Similarly, portions of the transposon and vector DNA can be hybridized to host cell genomic DNA to assess the number and uniqueness of insertions.

The utility of transposons in biological control experiments lies in their inherent marking of mutated genes coupled with the accompanying loss of the desired phenotype. Because the gene of interest can be located in the host cell genome, the transposon-tagged mutated gene from the phenotype-minus mutant can be cloned by marker rescue and used to identify the corresponding wild-type homologue from a genomic library of wild-type DNA. By mobilizing the wild-type gene back into the mutated strain, the lost phenotype may be restored, thus verifying the link between the particular gene with the phenotype of interest. This strategy has been used successfully to assess the role of various bacterial traits in biocontrol processes, including antibiotic biosynthesis (e.g., Fulton and Waddle, 1973; Georgakopoulos et al., 1994a,b), siderophore biosynthesis (e.g., Bull et al., 1994; Costa and Loper, 1994), hydrogen cyanide production (Voisard et al., 1989), adherence and exudate stimulant inactivation properties (Maloney et al., 1994; Nelson and Maloney, 1992), and various traits involved in root colonization (Lam et al., 1991).

Site-Directed Mutagenesis and Gene Replacement

Once genes involved in biocontrol processes have been identified, their role in biological control must be verified by further genetic manipulation. Cloned genes may be mutated by inserting a transposon carrying a selectable marker, or by deleting up to several nucleotide base pairs. They can be mobilized into the wild-type strain and exchanged for the wild-type allele by reciprocal crossovers (Ruvkin and Ausubel, 1981). If the gene is involved in biocontrol processes, the gene replacement should result in the loss or diminution of the biocontrol phenotype of the wild-type strain.

Reporter Genes

It may sometimes prove useful to characterize the expression or cellular localization of a gene impacting biocontrol processes by bringing the expression of a resident protein under the transcriptional control of a host-cell gene. Such reporter gene fusions — mutations that are fusion products of a resident gene to a foreign gene (Silhavy et al., 1984) — utilize the foreign protein to convert a substrate to a visible product. Reporter genes are commonly used to detect the expression of previously identified genes or to monitor the spatial distribution of cells. Depending on the particular construct, the expression of the foreign gene is then regulated by transcriptional and translational signals of the host gene. These types of constructs are particularly useful in determining factors that regulate the expression of biocontrol genes and identify mutations in specific classes of genes.

The more commonly used reporter genes include β-galactosidase (*lac*) (Kroos and Kaiser, 1984), β-glucuronidase (GUS) (Jefferson et al., 1986), alkaline phosphatase (*pho*A) (Manoil and Beckwith, 1985; Manoil et al., 1990), luciferase (*lux*) (Shaw and Kado, 1986), and ice nucleation reporters (Loper and Lindow, 1994). Many of these have been used in biological control studies. *Lac* fusions have been used to study the expression of antibiotic genes by *Pseudomonas fluorescens* suppressive to *Pythium ultimum* (Gutterson et al., 1988; Gutterson et al., 1986) and in tracking populations of *Pseudomonas* antagonists in soils and on plant roots (Drahos et al., 1986; Lam et al., 1991). Luciferase genes have also been used to study root colonization by *E. cloacae* in natural systems (Fravel et al., 1990). However, some constructs, particularly *pho*A, can be used directly in mutant library construction and screening to aid in the isolation of genes whose products are localized in different portions of a cell (Manoil et al., 1990). This reporter has also been used in biological control studies (Nelson and Maloney, 1992).

Transgenic Fungal Biocontrol Agents

One of the major factors hampering progress in understanding molecular mechanisms of biocontrol in fungi has been the limited availability of adequate gene transfer technologies. Up until recently, one of the more common methods of directed genetic recombination in biocontrol fungi had been protoplast fusion. This technique has been used extensively with species of *Trichoderma* (Pe'er and Chet, 1990; Stasz and Harman, 1990; Stasz et al., 1988) as an effective means of combining desirable phenotypes from different strains into one thallus. Pe'er and Chet (1990) demonstrated that, from the fusion of protoplasts from two relatively nonsuppressive auxotrophic strains of *T. harzianum*, prototrophic progeny could be obtained that were highly effective in suppressing Rhizoctonia damping-off of cotton. Additionally, these fusion progeny were more effective than the parental strains in overgrowing colonies of *S. rolfsii, R. solani,* and *P. aphanidermatum*. In addition to enhanced biological control activity, fusion progeny can be selected that are highly rhizosphere-competent (Harman, 1992; Sivan and Harman, 1991) and can enhance plant growth (Harman, 1989).

Recently, numerous transformation systems have been described for *Trichoderma* (Goldman et al., 1990; Herrera-Estrella et al., 1990; Lorito et al., 1993b; Pentila et al., 1987; Sivan et al., 1992) and *Gliocladium* (Lorito et al., 1993b; Ossana and Mischke, 1990; Thomas and Kennerley, 1989). These developments, along with a greater use of fungal transposon, will, for the first time, provide more precise methods of genetic disruption and complementation analysis and allow investigators to more adequately assess the role of various traits in biological control processes.

MECHANISMS OF BIOLOGICAL CONTROL OF *PYTHIUM* DISEASES BY *ENTEROBACTER CLOACAE*

Enterobacter cloacae is an effective biological protectant against infection from soilborne plant pathogens (Elad and Baker, 1985; Fravel et al., 1990; Hadar et al., 1983; Harman and Hadar, 1983; Howell et al., 1988; Lorito et al., 1993a; Lynch et al., 1991; Nelson, 1988; Nelson et al., 1986; Nelson and Craft, 1991; Nelson and Maloney, 1992; Roberts et al., 1992; Roberts et al., 1994; Sneh et al., 1984; Taylor et al., 1985). This organism is particularly effective in suppressing diseases incited by *Pythium* species (Hadar et al., 1983; Harman and Hadar, 1983; Harman and Nelson, 1994; Howell et al., 1988; Lynch et al., 1991; Nelson, 1988; Nelson, 1992; Nelson et al., 1986; Nelson and Maloney, 1992; Roberts et al., 1994; Taylor et al., 1985), but it is also effective against a number of other pathogenic fungi (Harman and Taylor, 1988; Lorito et al., 1993a; Sneh et al., 1984; Wilson et al., 1987; Wisniewski et al., 1989), including *Sclerotinia homoeocarpa* (Nelson and Craft, 1991).

The precise mechanisms of pathogen and disease suppression by *E. cloacae* are as yet unknown, although a number of traits have been empirically related to the suppression of seed and seedling rots caused by *Pythium ultimum* (Howell et al., 1988; Lorito et al., 1993a; Maloney and Nelson, 1992; Maloney and Nelson, 1994; Nelson et al., 1986; Nelson and Maloney, 1992; Roberts et al., 1992; Roberts et al., 1994; Trutmann and Nelson, 1992; Wilson et al., 1987; Wisniewski et al., 1989). To date, however, no conclusive results point to one major mechanism by which *E. cloacae* suppresses diseases caused by *Pythium* species.

One of the more conspicuous traits of *E. cloacae* in its interaction with *Pythium* spp. is its ability to adhere to hyphae. Empirical relationships between adherence of *E. cloacae* to hyphae of *P. ultimum* and biological control properties in the bacterium have been established (Nelson et al., 1986). However, pretreating cell suspensions of *E. cloacae* with various mono-, di-, and trisaccharides, as well as certain amino sugars, or α-linked glucosides, prevents cells from attaching to intact hyphae or agglutinating hyphal fragments. Addition of these same sugars also eliminates the ability of cells to inhibit fungal growth (Howell et al., 1988; Nelson et al., 1986; Wisniewski et al., 1989). On the other hand, pretreatment of cells with certain other monosaccharides, methylated sugars, or α-linked glucosides does not interfere with the ability of *E. cloacae* to attach to hyphae and inhibit the growth of *P. ultimum* (Nelson et al., 1986). The same carbohydrates that block binding of *E. cloacae* to *P. ultimum* hyphae also block ammonia production by *E. cloacae* (Howell et al., 1988) and allow cells to disperse through water films adjacent to *P. ultimum* hyphae (Maloney and Nelson, unpublished). These results suggest that bacterial adherence to hyphal cell walls might involve the binding of a fimbrial adhesion to specific sugar residues, possibly glucosides, that are associated with the fungal cell wall.

Microscopic studies of the interactions between bacteria and live mycelium have revealed bacterial cells clustered primarily around hyphal tips that are devoid of organized and streaming cytoplasm, around tips that are sealed off from their hyphae by septa, or along hyphae that contain cytoplasm that is no longer streaming and is withdrawn from the mycelial cell wall (Maloney and Nelson, unpublished). Only rarely are bacteria clustered in large numbers along hyphae that have actively streaming cytoplasm. These observations have led us to speculate about the nature of the *Pythium* cellular changes occurring as a result of the close interaction of *E. cloacae* with hyphal tips. It is unclear from our preliminary experiments whether *E. cloacae* cells bind preferentially to regions of hyphae devoid of cytoplasm, or whether binding of *E. cloacae* results in the observed loss of cytoplasm and hence viability. Although the empirical relationships between bacterial adherence and biological control activity appear to be firm, mutational analysis of these interactions has not supported the role of *E. cloacae* adherence properties in biological activity (Nelson and Maloney, 1992).

An understanding of biological control mechanisms begins with an understanding of the host-pathogen interaction targeted for control. One of the more important aspects of *Pythium* diseases of plants is that *Pythium* species respond extremely rapidly to germinating seeds and growing roots (Fukui et al., 1994a,b,c; Lifshitz et al., 1986; Loper, 1988; Nelson, 1987; Nelson, 1990; Nelson et al., 1986; Osburn and Schroth, 1988; Osburn et al., 1989; Stanghellini and Hancock, 1971; Xi et al., 1995). Since *Pythium* spp. are highly dependent on exudate molecules to initiate these rapid responses to and infection of plants (Nelson, 1990), microbial interference with the production and activity of such stimulatory molecules could be an effective mechanism of biological control of *Pythium* diseases.

Over the past few years we have been exploring the possibility of such a mechanism operating with *E. cloacae* and its suppression of *Pythium* diseases. Our previous research with other crop plants has shown that *E. cloacae* and other seed-applied rhizobacteria can utilize seed exudate from a variety of plant species as a sole carbon and energy source and, at the same time, rapidly reduce the stimulatory activity of exudate to *P. ultimum* sporangia (van Dijk, 1995). Depending on the cell density, this inactivation of exudate can occur as rapidly as 2–4 hr.

Other previous work in our laboratory indicated that unsaturated long chain fatty acids (LCFA) found in plant exudates were the primary molecules responsible for the elicitation of *Pythium* responses to plants. From analysis of these exudate fatty acids, we have found linoleic acid to be the most abundant unsaturated fatty acid found in exudates from a number of plant species, including creeping bentgrass (Penncross) and perennial ryegrass (All*Star) (Ruttledge and Nelson, 1996; Nelson and Ruttledge, unpublished).

There is now growing evidence that disruption of host-pathogen signalling by biocontrol organisms could potentially play an important role in disease suppression (Ahmad and Baker, 1988; Elad and Chet, 1987; Fukui et al., 1994; Gorecki et al., 1985; Maloney et al., 1994; Nelson, 1990; Paulitz, 1991; Paulitz et al., 1992). Therefore, our efforts have focussed on this question relative to the biological control of *Pythium* diseases of creeping bentgrass. We attempted to determine if the metabolism of unsaturated LCFA can be correlated with the inactivation of seed exudate stimulatory activity and whether this inactivation is related to the biological control of *Pythium* diseases of creeping bentgrass.

In an initial survey of bacteria recovered from seeds of various plant species, we found that, unlike our studies with cotton exudates, few of these bacterial strains could reduce the stimulatory activity of perennial ryegrass exudate within 24 hr to levels capable of inducing less than 30% *Pythium* sporangium germination (Table 5.2). However, of those strains with activity, the majority were strains of *Enterobacter cloacae*. Similarly, few strains were capable of inactivating the stimulatory activity of linoleic acid. However, there were good correlations between the ability of *E. cloacae* strains to inactivate linoleic acid and their ability to protect creeping bentgrass from *Pythium* damping-off. Despite the poor correlation between linoleic acid inactivation and biological control activity among the other genera and species, 11 of 18 strains were suppressive to *Pythium* damping-off of creeping bentgrass. This reflects the different mechanisms of biological control among *Pseudomonas* species as compared with *E. cloacae*.

Table 5.2. Reduction of the Stimulatory Activity of Perennial Ryegrass Seed Exudate and Linoleic Acid and Biological Control of *Pythium* Damping-Off of Creeping Bentgrass by Seed-Associated Rhizobacteria.

		% Germinated Sporangia		
Bacterium	**Source**	**Ryegrass Exudate[a]**	**Linoleic Acid[b]**	**Creeping Bentgrass Seedling Stand[c]**
Acinetobacter calcoaceticus AN3	Ryegrass	10.7*	0.0*	75.0*
Enterobacter cloacae MN5	Zucchini	4.8*	75.0	41.7
E. cloacae MN9	Zucchini	8.6*	13.1*	100.0*
E. cloacae EcCT-501	Cotton	0.0*	0.6*	83.3*
E. cloacae EN1	K. bluegrass	25.9*	90.1	58.3
Pantoea agglomerans IN2	Alfalfa	94.5	16.2*	58.3
Pseudomonas fragi CN1	Onion	77.1	75.9	58.3
P. cichorii CN2	Onion	50.0*	54.3*	25.0
P. corrugata DN3	K. bluegrass	89.4	79.4	100.0*
P. corrugata EN11	C. bentgrass	98.2	59.1	58.3
P. fluorescens IN4	Muskmelon	90.2	23.4*	75.0*
P. fluorescens AN4	Ryegrass	0.0*	82.3	100.0*
P. fluorescens FN3	C. bentgrass	70.2	23.9*	91.7*
P. fluorescens HN11	Alfalfa	99.4	77.5	66.7

Table 5.2. Reduction of the Stimulatory Activity of Perennial Ryegrass Seed Exudate and Linoleic Acid and Biological Control of *Pythium* Damping-Off of Creeping Bentgrass by Seed-Associated Rhizobacteria (Continued).

		% Germinated Sporangia		
Bacterium	**Source**	**Ryegrass Exudate[a]**	**Linoleic Acid[b]**	**Creeping Bentgrass Seedling Stand[c]**
P. fulva KN4	Tomato	—	44.5*	100.0*
P. putida AN3	Ryegrass	42.1*	77.5	83.3*
P. putida DN2	K. bluegrass	41.6*	64.0	100.0*
P. putida IN8	Muskmelon	99.3	53.2*	100.0*
Noninoculated control	—	88.1	93.0	100.0*
Untreated (*Pythium* control)	—	NA[d]	NA	25.0

[a] Inactivation of stimulatory activity of exudates by bacterial isolates was determined after 24 hr growth on 4-hr-old perennial ryegrass (All*Star) seed exudate. Numbers followed by an asterisk (*) are significantly (P=0.05) different from the noninoculated control according to the LSD test.

[b] Inactivation of stimulatory activity of linoleic acid solutions by bacterial isolates was determined after 24 hr growth on M9 medium containing 0.2 mL/mL sodium linoleate. Numbers followed by an asterisk (*) are significantly (P=0.05) different from the noninoculated control according to the LSD test.

[c] Bentgrass seedling stands were calculated 10 days after sowing as a percentage of the remaining healthy seedlings. Numbers followed by an asterisk (*) are significantly (P=0.05) different from the untreated (*Pythium*-inoculated) control according to the LSD test.

[d] NA = not applicable.

We have taken a multifaceted approach involving physiological, biochemical, and molecular studies to more closely investigate the relationship between fatty acid metabolism and biological control processes. Wild-type strains of *E. cloacae* can utilize a variety of both unsaturated and saturated LCFA as sole carbon and energy sources but grow very poorly, if at all, on medium-chain length (C_7–C_{11}) fatty acids (van Dijk, 1995). In addition, *E. cloacae* is capable of eliminating the stimulatory activity of unsaturated LCFA to *P. ultimum* sporangia in as little as 12 hr (Figure 5.1).

Much of our knowledge of fatty acid metabolism in gram-negative bacteria comes from physiological, biochemical, and genetic studies in *Escherichia coli* (Black and DiRusso, 1994). The critical first step in the degradation of fatty acids in enteric bacteria is the transport of these molecules into the cell. Three genes, designated *fadL*, *fadD*, and *tsp* (Azizan and Black, 1994; Black et al., 1992; Black et al., 1985) encode an outer-membrane protein (FadL), an acyl-CoA-synthetase (FadD), and a periplasmic cotransporter protein (Tsp) involved in transport activation, respectively, that are all involved in the transport of fatty acids to the cytoplasm (Black and DiRusso, 1994). FadL is essential for both the binding of exogenous LCFA and its transport into the periplasm (Black, 1991; Black et al., 1987; Ginsburgh et al., 1984; Kumar and Black, 1993). Its expression is induced by the presence of LCFA in the environment. Medium-chain fatty acids (MCFA) also can be transported by FadL. However, if FadL is nonfunctional or uninduced, MCFA can diffuse through the outer membrane. LCFA and MCFA transit the periplasmic space by an as yet undescribed mechanism, but this transport is potentiated by Tsp and may be involved in ligand release from FadL (Azizan and Black, 1994). Fatty acids are transported unidirectionally across the inner membrane and are concomitantly activated to their acyl-CoA thioester derivatives by FadD. This is the terminal step in LCFA and MCFA uptake and the initial step in the β-oxidation pathway, guaranteeing irreversible LCFA and MCFA uptake by the cell.

The β-oxidation pathway has been well studied in *E. coli* (Black and DiRusso, 1994), and can be briefly summarized as follows. The utilization of LCFA (C_{12} and greater) by *E. coli* requires the FadR-mediated derepression of the fatty acid transport system described above, and the β-oxidation enzymes. In the oxidation of saturated fatty acids, the enzymes acyl-CoA synthetase, acyl-CoA dehydrogenase, enoyl-CoA hydratase, 3-hydroxyacyl-CoA dehydrogenase, and 3-ketoacyl-CoA thiolase are required for successive two-carbon shortening of the fatty acid chain, giving rise to acetyl CoA. For unsaturated fatty acids, the additional enzymes 3-hydroxyacyl-CoA epimerase and *cis*-Δ^3-*trans*-Δ^2-enoyl-CoA isomerase are required.

The genes encoding all of the β-oxidation enzymes have been cloned and mapped to different sites within the *E. coli* genome. The genes responsible for the transport, acylation, and β-oxidation of LCFA comprise the *fad* regulon, which is induced by LCFA, but not MCFA, even though MCFA serve as substrates for FadD and the other β-oxidation enzymes. (Uptake and metabolism of short-chain fatty acids (C_3–C_6) are controlled by a distinct set of genes that

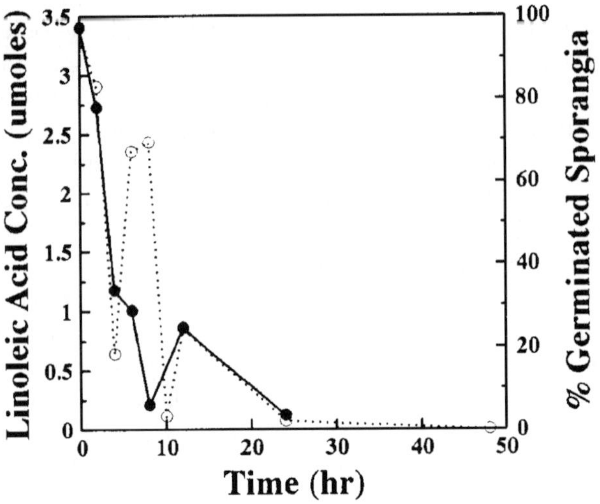

Figure 5.1. Metabolism of linoleic acid by *E. cloacae* strain EcCT-501. Solid line represents stimulatory activity; dotted line represents linoleic acid concentration.

comprise the *ato* regulon (Nunn, 1987)). Genes encoding the β-oxidation enzymes, *fadE*, *fadAB*, *fadH*, together with *fadL* and *fadD*, are all transcriptionally regulated by the product of the *fadR* gene (FadR protein) (Black and DiRusso, 1994). In addition, transcription of the fatty acid biosynthetic gene, *fabA*, is coregulated by FadR. FadR is a DNA-binding protein that represses the *fad* regulon but activates the *fabA* operon (Henry and Cronan, 1991; Henry and Cronan, 1992). In the presence of long-chain acyl-CoA fatty acid derivatives, FadR releases the *fadAB* and *fadL* promoter, as well as the *fabA* promoters, but with opposite effects: transcription of the *fad* genes is derepressed, while transcription of *fabA* is repressed (DiRusso et al., 1993). The activity of *fadR* is highly repressed by glucose (Nunn, 1987).

In our studies, we have taken advantage of the availability of mini-Tn*5* constructs (de Lorenzo et al., 1990) to generate mutants deficient in fatty acid metabolism so that relationships between this function and biological control activity can be observed. In initial studies, we were able to generate a number of mutants of *E. cloacae* strain EcCT-501 that were no longer suppressive to *P. ultimum* in cucumber and creeping bentgrass assays (Maloney et al., 1994; Nelson and Maloney, 1992). It turned out, however, that these were largely mutations in genes encoding TCA (tricarboxylic acid) and DCA (dicarboxylic) cycle enzymes.

One mutant in particular, V-58, is a malate dehydrogenase (*mdh*) mutant (Maloney et al., 1994; Maloney and Nelson, unpublished). Although *mdh* is not directly involved in fatty acid degradation, the high degree of regulation of fatty acid catabolism as observed in *E. coli* make *mdh* and other DCA enzymes potentially important regulators of fatty acid transport processes (Nunn, 1986).

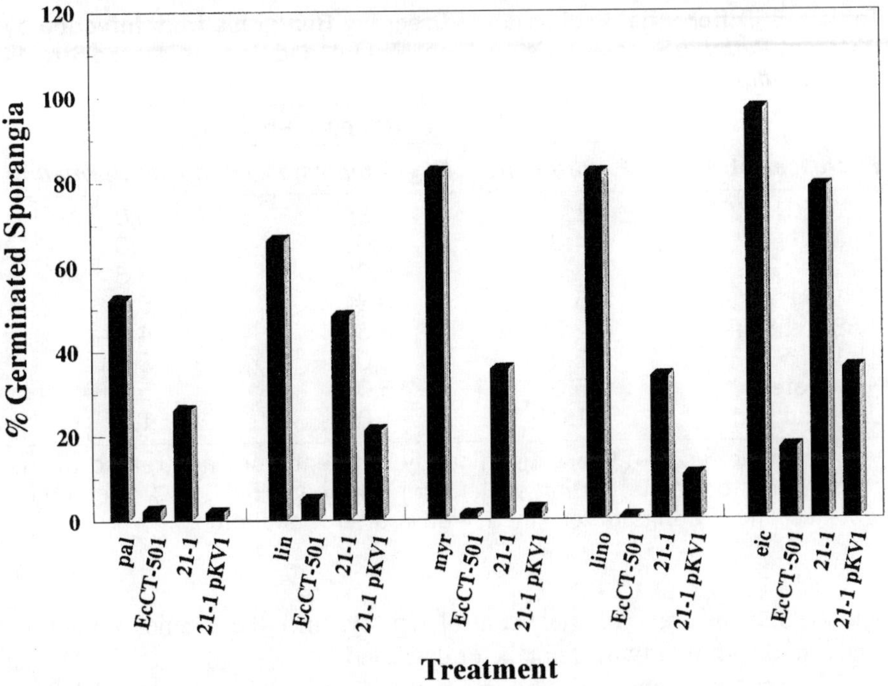

Figure 5.2. Inactivation of selected unsaturated fatty acids by *E. cloacae* strain EcCT-501 (wild-type), 21-1 (a biocontrol-minus mutant) and 21-1 (pKV1) (complemented mutant with a biocontrol-positive phenotype) after 24 hr growth. Cell-free filtrates were assayed for *Pythium ultimum* sporangium germination. pal=palmitoleic acid, lin=linoleic acid, myr=myristoleic acid, lino=linolenic acid, eic=eicosapentaenoic acid.

Another mutant (designated as strain 21-1) lacks the ability to grow on linoleic acid and is incapable of inactivating its stimulatory activity and that of other unsaturated fatty acids (Figure 5.2). Furthermore, this mutant fails to suppress *Pythium* infections of creeping bentgrass seedling (Table 5.3). Although some residual suppression of *P. graminicola* was observed with 21-1, this was significantly and substantially different from uninoculated seedlings and is not believed to be a significant level of biological control activity. Complementing strain 21-1 as well as V58 with the cosmids pV58K and pKV1, respectively, partially restored biological control activity to that of wild-type levels — levels that did not differ from uninoculated seedlings. Sequence analysis of the mutated gene in strain 21-1 indicated a high homology with an *E. coli* ATPase, an enzyme involved in oxidative phosphorylation and other important metabolic processes in the cell. Although V58 and 21-1 demonstrate the utility of generating mutants deficient in fatty acid metabolism and biological control activity, they reveal little about the specific link between fatty

Table 5.3. Differential Protection of Creeping Bentgrass from Infection by Different *Pythium* Species by Wild-Type, Mutant, and Complemented Strains of *Enterobacter Cloacae.*

	Disease Rating (1–5 Scale)[a]		
E. cloacae strain	*P. ultimum*	*P. graminicola*	*P. aphanidermatum*
EcCT-501 (WT)	1.8*	1.3*	2.0*
V-58	4.8	5.0	5.0
3-1	4.0	5.0	3.3*
21-1	5.0	3.8*	5.0
V-58(pV58K)	2.3*	2.8*	4.3*
21-1(pKV1)	2.0*	2.0*	3.8*
Nontreated	5.0	5.0	5.0
Uninoculated	1.0*	1.0*	1.0*

[a] Means followed by (*) are significantly different from nontreated plants according to T-tests. Rating scale: 1 = healthy and 5 = 100% unemerged or necrotic. Ratings determined 7 days after inoculation.

acid metabolism and biological control activity, since these genes are not directly involved in fatty acid uptake or degradation.

More recently, another fatty acid mutant (strain 3-1) was obtained. This mutant fails to grow on media containing linoleic acid as a sole carbon source, but grows well on a minimal media containing succinate. This selection protocol was chosen to avoid selecting mutants with disrupted TCA and DCA cycle enzymes. As with mutants V58 and 21-1, mutant 3-1 is unable to reduce the stimulatory activity of linoleic acid. But it very slowly reduces the stimulatory activity of seed exudate (Figure 5.3) and fails to protect bentgrass seedlings from infection by *Pythium* species (Table 5.3). Subsequent complementation and sequence analysis has revealed that the mutation in strain 3-1 is in the *fadAB* operon, which encodes five structural genes central to the β-oxidation of fatty acids. While this mutant is debilitated in its ability to catabolize linoleic acid, it is not clear whether this mutation represents deficiencies in linoleic acid transport or in linoleic acid utilization. Nonetheless, this mutant provides us with a more direct link between the two phenotypes of interest.

As a more direct approach of investigating fatty acid transport in *E. cloacae*, we are using cloned *fadL* and *fadD* genes from *E. coli* as probes for homologous transport genes in *E. cloacae.* In our initial experiments, we have found a significant homology with *fadL,* whereas the work with *fadD* is still in progress. These genes, along with the FadAB genes, will provide a more direct means of assessing the role of fatty acid uptake and metabolism in biological control processes with *E. cloacae.*

Currently, there are at least two explanations for fatty acid stimulant inactivation by *E. cloacae* and how this inactivation might be related to biological control processes. First, it is very likely that LCFA uptake and the β-oxidation pathway are centrally involved in removing or reducing the concentration of

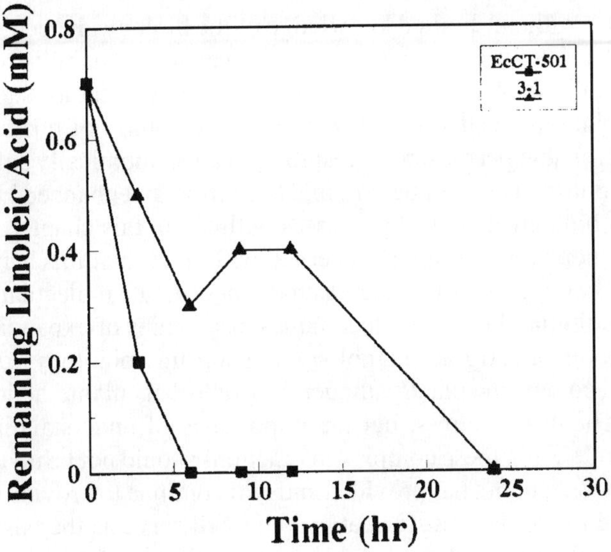

Figure 5.3. Inactivation of stimulants present in creeping bentgrass seed exudate by *E. cloacae* strains EcCT-501 and a *fadAB* mutant strain 3-1. Cell-free exudate filtrates were used to assess germination of *P. ultimum* sporangia.

linoleic acid and possibly other unsaturated fatty acid germination stimulants in plant exudates. By simply removing the fatty acid stimulant from solution through normal uptake mechanisms, the molecule becomes unavailable to *Pythium*. In this case, we predict that genes analogous to *fadL*, *fadD*, and *tsp* in *E. coli*, each of which encodes a specific single receptor/transport protein for LCFA (Azizan and Black, 1994; Nunn, 1987), are responsible for initial events in this process. Second, the biohydrogenation of linoleic acid to stearic acid or other saturated derivatives that lack stimulatory activity could result in the reduction or elimination of stimulatory activity in exudate or fatty acid solutions. Our previous work clearly demonstrated that saturated fatty acids contain little or no stimulatory activity (Ruttledge and Nelson, 1996). Preliminary evidence in our laboratory indicates that the biohydrogenation of linoleic acid to stearic acid by *E. cloacae* strain EcCT-501 could occur (Ruttledge and Nelson, unpublished). Biohydrogenation of unsaturated fatty acids has been described for various bacteria, but to our knowledge this process has not been investigated in bacteria associated with plants under aerobic conditions.

CONCLUSIONS

Gaps in our understanding of the processes by which plant-associated rhizobacteria interfere with activities of seed- and root-infecting fungal pathogens, coupled with our limited knowledge of the biology and ecology of target pathogens, have contributed in an important way to our inability to predict and manipulate microbial biological control systems. The future success of manipulating plant-associated rhizobacteria for the control of turfgrass diseases will require reliable performance, resulting in an economically sufficient level of disease control. This will be possible only from an enhanced knowledge of the types of biological control processes outlined in this chapter.

It is my hope that research concerned with microbial mechanisms of disease control in turfgrasses will increasingly incorporate molecular methodologies with traditional laboratory techniques as a means of expanding the more traditional tools of turfgrass pathology and soil microbiology. Questions related to soil ecology and plant/pathogen/biocontrol agent interactions are complex and difficult to address, but the importance of understanding the basic principles underlying these complex interactions should not be underestimated, since this understanding has provided and will continue to provide useful genes for plant and microbial transformations and will serve as the basis for developing more effective and predictable biological control organisms and disease-resistant plants in the future.

REFERENCES

Ahl, P., C. Voisard, and G. Defago. 1986. Iron bound siderophores, cyanic acid, and antibiotics involved in suppression of *Thielaviopsis basicola* by a *Pseudomonas fluorescens* strain. *J. Phytopath.,* 116:121–134.

Ahmad, J.S. and R. Baker. 1987a. Competitive saprophytic ability and cellulolytic activity of rhizosphere-competent mutants of *Trichoderma harzianum. Phytopathology,* 77:358–362.

Ahmad, J.S. and R. Baker. 1987b. Rhizosphere competence of *Trichoderma harzianum. Phytopathology,* 77:182–189.

Ahmad, J.S. and R. Baker. 1988. Implications of rhizosphere competence of *Trichoderma harzianum. Can. J. Microbiol.,* 34:229–234.

Azizan, A. and P.N. Black. 1994. Use of transposon Tn*phoA* to identify genes for cell envelope proteins of *Escherichia coli* required for long-chain fatty acid transport: the periplasmic protein Tsp potentiates long-chain fatty acid transport. *J. Bacteriol.,* 176:6653–6662.

Baldwin, N.A., A.L. Capper, and D.J. Yarham (Eds.). 1991. Evaluation of biological agents for the control of take-all patch (*Gaeumannomyces graminis*) of fine turf. In *Developments in Agricultural and Managed-Forest Ecology.* A.B.R.e.a. Beemster. Elsevier Science Publishers, Amsterdam.

Becker, J.O. and R.J. Cook. 1988. Role of siderophores in suppression of pythium species and production of increased growth response of wheat by fluorescent pseudomonads. *Phytopatholgy,* 78:778–782.

Berg, C.M. and D.E. Berg. 1987. Use of transposable elements and maps of known insertions. In *Escherichia coli and Salmonella typhimurium — Cel-*

lular and Molecular Biology. F.C. Neidhardt (Ed.). American Society for Microbiology, Washington, DC.

Berg, D.E. and C.M. Berg. 1983. The prokaryotic transposable element Tn*5*. *Bio/Technology,* 1:417–435.

Black, P.N. 1991. Primary sequence of the *Escherichia coli fadL* gene encoding an outer membrane protein required for long-chain fatty acid transport. *J. Bacteriol.,* 173:435–442.

Black, P.N. and C.C. DiRusso. 1994. Molecular and biochemical analyses of fatty acid transport, metabolism, and gene regulation in *Escherichia coli. BBA,* 1210:123–145.

Black, P.N., C.C. DiRusso, A.K. Metzger, and T.L. Heimert. 1992. Cloning, sequencing, and expression of the *fadD* gene of *Escherichia coli* encoding acyl coenzyme A synthetase. *J. Biol. Chem.,* 267:25513–25520.

Black, P.N., S.F. Kianian, C.C. DiRusso, and W.D. Nunn. 1985. Long-chain fatty acid transport in *Escherichia coli*: Cloning, mapping, and expression of the *fadL* gene. *J. Biol. Chem.,* 260:1780–1789.

Black, P.N., B. Said, C.R. Ghosn, J.V. Beach, and W.D. Nunn. 1987. Purification and characterization of an outer membrane-bound protein involved in long-chain fatty acid transport in *Escherichia coli. J. Biol. Chem.,* 262:1412–1419.

Bol, J.F., A.S. Buchel, M. Knoester, T. Baldwin, L.C. Van Loon, and H.J.M. Linthorst. 1996. Regulation of the expression of plant defense genes. *Plant Growth Regulation,* 18:87–91.

Broglie, K., I. Chet, M. Holliday, R. Cressman, P. Biddle, S. Knowlton, C.J. Mauvais, and R. Broglie. 1991. Transgenic plants with enhanced resistance to the fungal pathogen *Rhizoctonia solani. Science,* 254:1194–1197.

Bull, C.T., C.A. Ishimaru, and J.E. Loper. 1994. Two genomic regions involved in catechol siderophore production in *Erwinia carotovora. Appl. Environ. Microbiol.,* 60:662–669.

Burkhead, K.D., D.A. Schisler, and P.J. Slininger. 1994. Pyrollnitrin production by biological control agent *Pseudomonas cepacia* B37w in culture and in colonized wounds of potatoes. *Appl. Environ. Microbiol.,* 60:2031–2039.

Burpee, L.L. 1994. Interactions among low-temperature-tolerant fungi: prelude to biological control. *Can. J. Pl. Path.,* 16:247–250.

Burpee, L.L. and L.G. Goulty. 1984. Suppression of brown patch disease of creeping bentgrass by isolates of non-pathogenic *Rhizoctonia* spp. *Phytopathology,* 74:692–694.

Burpee, L.L., L.M. Kaye, L.G. Goulty, and M.B. Lawton. 1987. Suppression of gray snow mold on creeping bentgrass by an isolate of *Typhula phacorrhiza. Plant Dis.,* 71:97–100.

Buyer, J.S., M.G. Kratzke, and L.J. Sikora. 1994. Microbial siderophores and rhizosphere ecology. In *Biochemistry of Metal Micronutrients in the Rhizosphere.* J.A. Manthey, D.E. Crowley, and D.G. Luster (Eds.). CRC Press, Inc., Boca Raton, FL. 67–80 pp.

Buysens, S., K. Heungens, J. Poppe, and M. Höfte. 1996. Involvement of pyochelin and pyoverdin in suppression of *Pythium*-induced damping-off of tomato by *Pseudomonas aeruginosa* 7NSK2. *Appl. Environ. Microbiol.,* 62:865–871.

Carroll, H., Y. Moenneloccoz, D.N. Dowling, and F. Ogara. 1995. Mutational disruption of the biosynthesis genes coding for the antifungal metabolite 2,4-diacetylphloroglucinol does not influence the ecological fitness of *Pseudomonas fluorescens* f113 in the rhizosphere of sugarbeets. *Appl. Environ. Microbiol.,* 61:3002–3007.

Chernin, L., A. Brandis, Z. Ismailov, and I. Chet. 1996. Pyrrolnitrin production by an *Enterobacter agglomerans* strain with a broad spectrum of antagonistic activity towards fungal and bacterial phytopathogens. *Curr. Microbiol.,* 32:208–212.

Chernin, L., Z. Ismailov, S. Haran, and I. Chet. 1995. Chitinolytic *Enterobacter agglomerans* antagonistic to fungal plant pathogens. *Appl. Environ. Microbiol.,* 61:1720–1726.

Chet, I. 1987. *Trichoderma* — Application, mode of action, and potential as a biocontrol agent of soilborne plant pathogenic fungi. In *Innovative Approaches to Plant Disease Control.* I. Chet (Ed.). John Wiley & Sons, NY.

Claydon, N. 1987. Antifungal alkyl pyrones of *Trichoderma harzianum. Transactions of the British Mycological Society,* 88:503–513.

Cornelissen, B.J.C. and L.S. Melchers. 1993. Strategies for control of fungal diseases with transgenic plants. *Plant Physiol.,* 101:709–712.

Costa, J.M. and J.E. Loper. 1994. Characterization of siderophore production by the biological control agent *Enterobacter cloacae. Molecular Plant — Microbe Interactions,* 7:440–448.

Curl, E.A. and B. Truelove. 1986. *The Rhizosphere.* Advanced Series in Agricultural Sciences, Vol. 15. B. Yaron (Ed.). Springer-Verlag, NY. 288 pp.

Daboussi, M.J. and T. Langin. 1994. Transposable elements in the fungal plant pathogen *Fusarium oxysporum. Genetica,* 93:49–59.

de Lorenzo, V., M. Herrero, U. Jakubzik, and K.N. Timmis. 1990. Mini-Tn*5* transposon derivatives for insertion mutagenesis, promoter probing, and chromosomal insertion of cloned DNA in gram-negative eubacteria. *J. Bacteriol.,* 172:6568–6572.

Deacon, J.W. 1973. Factors affecting occurrence of the *Ophiobolus* patch disease of turf and its control by *Phialophora radicicola. Plant Pathol.,* 22:149–155.

DiRusso, C.C., A.K. Metzger, and T.L. Heimert. 1993. Regulation of transcription of genes required for fatty acid transport and unsaturated fatty acid biosynthesis in *Escherichia coli* by FadR. *Molec. Microbiol.,* 7:311–322.

Drahos, D.J., B.C. Hemming, and S. McPherson. 1986. Tracking recombinant organisms in the environment: β-galactosidase as a selectable non-antibiotic marker for fluorescent pseudomonads. *Biotechnology,* 4:439–444.

Elad, Y. and R. Baker. 1985. The role of competition for iron and carbon in suppression of chlamydospore germination of *Fusarium* spp. by *Pseudomonas* spp. *Phytopathology,* 75:1053–1059.

Elad, Y. and R. Barak. 1983. Possible role of lectins in mycoparasitism. *J. Bacteriol.,* 154:1431–1435.

Elad, Y. and I. Chet. 1987. Possible role of competition for nutrients in biocontrol of Pythium damping-off by bacteria. *Phytopathology,* 77:190–195.

Fenton, A.M., P.M. Stephens, J. Crowley, M. Ocallaghan, and F. Ogara. 1992. Exploitation of gene(s) involved in 2,4-diacetylphloroglucinol biosynthesis to confer a new biocontrol capability to a *Pseudomonas* strain. *Appl. Environ. Microbiol.,* 58:3873–3878.

Fravel, D.R., R.D. Lumsden, and D.P. Roberts. 1990. *In situ* visualization of the biocontrol rhizobacterium *Enterobacter cloacae* with bioluminescence. *Plant Soil,* 125:233–238.

Fuchs, R.L., S.A. McPherson, and D.J. Drahos. 1986. Cloning of a *Serratia marcescens* gene encoding chitinase. *Appl. Environ. Microbiol.,* 51:504–509.

Fukui, R., G.S. Campbell, and R.J. Cook. 1994a. Factors influencing the incidence of embryo infection by *Pythium* spp. during germination of wheat seeds in soils. *Phytopathology,* 84:695–702.

Fukui, R., E.I. Poinar, P.H. Bauer, M.N. Schroth, M. Hendson, X.L. Wang, and J.G. Hancock. 1994b. Spatial colonization patterns and interaction of bacteria on inoculated sugar beet seed. *Phytopathology,* 84:1338–1345.

Fukui, R., M.N. Schroth, M. Hendson, and J.G. Hancock. 1994c. Interaction between strains of pseudomonads in sugar beet spermospheres and their relationship to pericarp colonization by *Pythium ultimum* in soil. *Phytopathology,* 84:1322–1330.

Fulton, N.D. and B.A. Waddle. 1973. Searching for possible tolerance in cotton to seedling disease. *Arkansas Farm Research,* 9–10.

Georgakopoulos, D.G., M. Hendson, N.J. Panopoulos, and M.N. Schroth. 1994a. Analysis of expression of a phenazine biosynthesis locus of *Pseudomonas aureofaciens* PGS12 on seeds with a mutant carrying a phenazine biosynthesis locus ice nucleation reporter gene fusion. *Appl. Environ. Microbiol.,* 60:4573–4579.

Georgakopoulos, D.G., M. Hendson, N.J. Panopoulos, and M.N. Schroth. 1994b. Cloning of a phenazine biosynthetic locus of *Pseudomonas aureofaciens* PGS12 and analysis of its expression in vitro with the ice nucleation reporter gene. *Appl. Environ. Microbiol.,* 60:2931–2938.

Geremia, R.A., G.H. Goldman, D. Jacobs, W. Ardiles, S.B. Vila, M. van Montagu, and A. Herrera-Estrella. 1993. Molecular characterization of the proteinase-encoding gene, prb1, related to mycoparasitism by *Trichoderma harzianum. Molec. Microbiol.,* 8:603–613.

Giesler, L.J., G.Y. Yuen, and M.L. Craig. 1993. Evaluation of fungal antagonists against *Rhizoctonia solani* in tall fescue. *Biol. Cult. Tests Cont. Plant Dis.,* 7:123.

Ginsburgh, C.L., P.N. Black, and W.D. Nunn. 1984. Transport of long-chain fatty acids in *Escherichia coli*: Identification of a membrane protein associated with the *fadL* gene. *J. Biol. Chem.,* 259:8437–8443.

Glayzer, D.C., I.N. Roberts, D.B. Archer, and R.P. Oliver. 1995. The isolation of Ant*1*, a transposable element from *Aspergillus niger. Mol. Gen. Genet.,* 249:432–438.

Goldman, G.H., M. VanMontagu, and A. Herrera-Estrella. 1990. Transformation of *Trichoderma harzianum* by high-voltage electric pulse. *Current Genetics,* 17:169–174.

Goodman, D.M. and L.L. Burpee. 1991. Biological control of dollar spot disease on creeping bentgrass. *Phytopathology,* 81:1438–1446.

Gorecki, R.J., G.E. Harman, and L.R. Mattick. 1985. The volatile exudates from germinating pea seeds of different viability and vigor. *Can. J. Bot.,* 63:1035–1039.

Grebus, M.E., J.W. Rimelspach, and H.A.J. Hoitink. 1995. Control of dollar spot of turf with biocontrol agent-fortified compost topdressings. *Phytopathology,* 85:1166.

Grebus, M.E., M.E. Watson, and H.A.J. Hoitink. 1994. Biological, chemical, and physical properties of composted yard trimmings as indicators of maturity and plant disease suppression. *Compost Sci. Util.,* 2:57–71.

Gutterson, N., J.S. Ziegle, G.J. Warren, and T.J. Layton. 1988. Genetic determinants for catabolite induction of antibiotic biosynthesis in *Pseudomonas fluorescens* HV37a. *J. Bacteriol.,* 170:380–385.

Gutterson, N.I., T.J. Layton, J.S. Ziegle, and G.J. Warren. 1986. Molecular cloning of genetic determinants for inhibition of fungal growth by a fluorescent pseudomonad. *J. Bacteriol.,* 165:696–703.

Hadar, Y., G.E. Harman, A.G. Taylor, and J.M. Norton. 1983. Effects of pre-germination of pea and cucumber seeds and of seed treatment with *Enterobacter cloacae* on rots caused by *Pythium* spp. *Phytopathology,* 73:1322–1325.

Haran, S., H. Schickler, S. Pe'er, S. Logeman, A. Oppenheim, and I. Chet. 1993. Increased constitutive chitinase activity in transformed *Trichoderma harzianum. Biol. Contr.,* 3:101–108.

Harder, P.R. and J. Troll. 1973. Antagonism of *Trichoderma* spp. to sclerotia of *Typhula incarnata. Plant Dis. Rptr.,* 57:924–926.

Harman, G.E. 1989. Combining effective strains of *Trichoderma harzianum* and solid matrix priming to improve biological seed treatments. *Plant Dis.,* 73:631–637.

Harman, G.E. 1992. Development and benefits of rhizosphere competent fungi for biological control of plant pathogens. *J. Plant Nutrition,* 15:835–843.

Harman, G.E. and Y. Hadar. 1983. Biological control of *Pythium* species. *Seed Sci. Technol.,* 11:893–906.

Harman, G.E. and C.K. Hayes. 1993. The genome of biocontrol fungi: Modification and genetic components for disease management. In *Pest Management: Biologically-Based Technologies.* R.D. Lumsden and J.L. Vaughn (Eds.). American Chemical Society, Washington, DC.

Harman, G.E. and C.T. Lo. 1996. The first registered biological control product for turf disease: BioTrek 22G. *Turfgrass TRENDS,* 5(5):8–14.

Harman, G.E. and E.B. Nelson. 1994. Mechanisms of protection of seed and seedlings by biological seed treatments: Implications for practical disease control. In *Seed Treatment: Progress and Prospects, BCPC Monograph No. 57.* T. Martin, Ed.. British Crop Protection Council, Farnham, UK.

Harman, G.E. and A.G. Taylor. 1988. Improved seedling performance by integration of biological control agents at favorable pH levels with solid matrix priming. *Phytopathology,* 78:520–525.

Hayes, C.K., S. Klemsdal, M. Lorito, A. Di Pietro, C. Peterbauer, J. Nakas, A. Tronsmo, and G.E. Harman. 1994. Isolation and sequence of an endochitinase gene from a cDNA library of *Trichoderma harzianum. Gene,* 138:143–148.

Haygood, R.A. and A.R. Mazur. 1990. Evaluation of *Gliocladium virens* as a biocontrol agent for dollar spot on bermudagrass. *Phytopathology,* 80:435.

Haygood, R.A. and J.F. Walter. 1991. Biocontrol of brown patch of centipedegrass with *Gliocladium virens* in a growth chamber. *Biol. Cult. Tests Cont. Plant Dis.,* 5:86.

Henry, M.F. and J.E. Cronan. 1991. *Escherichia coli* transcription factor that both activates fatty acid synthesis and represses fatty acid degradation. *J. Molec. Biol.,* 222:843–849.

Henry, M.F. and J.E. Cronan. 1992. A new mechanism of transcriptional regulation: release of an activator triggered by a small molecule binding. *Cell,* 70:671–679.

Herrera-Estrella, A., G.H. Goldman, and M. VanMontagu. 1990. High-efficiency transformation system for the biocontrol agents, *Trichoderma* spp. *Molec. Microbiol.,* 4:839–843.

Hill, D.S., J.I. Stein, N.R. Torkewitz, A.M. Morse, C.R. Howell, J.P. Pachlatko, J.O. Becker, and J.M. Ligon. 1994. Cloning of genes involved in the synthesis of pyrrolnitrin from *Pseudomonas fluorescens* and role of pyrrolnitrin synthesis in biological control of plant disease. *Appl. Environ. Microbiol.,* 60:78–85.

Hodges, C.F., D.A. Campbell, and N. Christians. 1993. Evaluation of *Streptomyces* for biocontrol of *Bipolaris sorokiniana* and *Sclerotinia homoeocarpa* on the phylloplane of *Poa pratensis. J. Phytopath.,* 139:103–109.

Hodges, C.F., D.A. Campbell, and N. Christians. 1994. Potential biocontrol of *Sclerotinia homoeocarpa* and *Bipolaris sorokiniana* on the phylloplane of *Poa pratensis* with strains of *Pseudomonas* spp. *Plant Pathol.,* 43:500–506.

Homma, Y. 1984. Perforation and lysis of hyphae of *Rhizoctonia solani* and conidia of *Cochliobolus miyabeanus* by soil myxobacteria. *Phytopathology,* 74:1234–1239.

Homma, Y. and T. Suzui. 1989. Role of antibiotic production in suppression of radish damping-off by seed bacterization with *Pseudomonas cepacia. Ann. Phytopath. Soc. Japan,* 55:643–652.

Howell, C.R., R.C. Beier, and R.D. Stipanovic. 1988. Production of ammonia by *Enterobacter cloacae* and its possible role in the biological control of Pythium preemergence damping-off by the bacterium. *Phytopathology,* 78:1075–1078.

Howell, C.R. and R.D. Stipanovic. 1980. Suppression of *Pythium ultimum*-induced damping-off of cotton seedlings by *Pseudomonas fluorescens* and its antibiotic, pyoluteorin. *Phytopathology,* 70:712–715.

Howell, C.R. and R.D. Stipanovic. 1983. Gliovirin, a new antibiotic from *Gliocladium virens*, and its role in the biological control of *Pythium ultimum. Can. J. Microbiol.,* 29:321–324.

Howie, W., L. Joe, E. Newbigin, T. Suslow, and P. Dunsmuir. 1994. Transgenic tobacco plants which express the *chiA* gene from *Serratia marcescens* have enhanced tolerance to *Rhizoctonia solani. Transgenic Research,* 3:90–98.

Howie, W. and T.V. Suslow. 1986. Effect of antifungal compound biosynthesis on cotton root colonization and *Pythium* suppression by a strain of *Pseudomonas fluorescens* and its antifungal minus isogenic mutant. *Phytopathology,* 76:1069.

Howie, W.J. and T.V. Suslow. 1991. Role of antibiotic biosynthesis in the inhibition of *Pythium ultimum* in the cotton spermosphere and rhizosphere by *Pseudomonas fluorescens. Molec. Plant-Microbe Interact.,* 4:393–399.

Hubbard, J.P. and G.E. Harman. 1983. Effect of soilborne *Pseudomonas* spp. on the biological control agent, *Trichoderma hamatum,* on pea seeds. *Phytopathology,* 73:655–659.

James, D.W.J. and N.I. Gutterson. 1986. Multiple antibiotics produced by *Pseudomonas fluorescens* HV37a and their differential regulation by glucose. *Appl. Environ. Microbiol.,* 52:1183–1189,

Jefferson, R.A., S.M. Burgess, and D. Hirsch. 1986. β-glucuronidase from *Escherichia coli* as a gene-fusion marker. *Proc. Natl. Acad. Sci. USA,* 83:8447–8451.

Jones, J.D.G., K.L. Grady, T.V. Suslow, and J.R. Bedbrook. 1986. Isolation and characterization of genes encoding two chitinase enzymes from *Serratia marcescens. EMBO J.,* 5:467–473.

Keel, C. 1990. Pseudomonads as antagonists of plant pathogens in the rhizosphere: Role of the antibiotic 2,4-diacetylphloroglucinol in the suppression. *Symbiosis,* 9:327–341.

Keel, C., U. Schnider, M. Maurhofer, C. Voisard, J. Laville, U. Burger, P. Wirthner, D. Haas, and G. Défago. 1992. Suppression of root diseases by *Pseudomonas fluorescens* CHA0: Importance of the bacterial secondary metabolite 2,4-diacetylphloroglucinol. *Molec. Plant-Microbe Interact.,* 5:4–13.

Kloepper, J.W. and J. Leong. 1980. Pseudomonas siderophores: a mechanism explaining disease-suppressive soils. *Curr. Microbiol.,* 4:317–320.

Kobayashi, D.Y. and N.E.-H. El-Barrad. 1996. Selection of bacterial antagonists using enrichment cultures for the control of summer patch disease on Kentucky bluegrass. *Curr. Microbiol.,* 32:106–110.

Kobayashi, D.Y., M. Guglielmoni, and B.B. Clarke. 1995. Isolation of the chitinolytic bacteria *Xanthomonas maltophilia* and *Serratia marcescens* as biological control agents for summer patch disease of turfgrass. *Soil Biol. Biochem.,* 27:1479–1487.

Koby, S., H. Schickler, I. Chet, and A.B. Oppenheim. 1994. The chitinase encoding Tn7-based chiA gene endows *Pseudomonas fluorescens* with the capacity to control plant pathogens in soil. *Gene,* 147:81–83.

Kraus, J. and J.E. Loper. 1992. Lack of evidence for a role of antifungal metabolite production by *Pseudomonas fluorescens* Pf-5 in biological control of Pythium damping-off of cucumber. *Phytopathology,* 82:264–271.

Kraus, J. and J.E. Loper. 1995. Characterization of a genomic region required for production of the antibiotic pyoluteorin by the biological control agent *Pseudomonas fluorescens* Pf5. *Appl. Environ. Microbiol.,* 61:849–854.

Kroos, L. and D. Kaiser. 1984. Construction of Tn*5 lac*, a transposon that fuses *lacZ* expression to exogenous promoters, and its introduction into *Myxococcus xanthus*. *Proc. Natl. Acad. Sci. USA,* 81:5816–5820.

Kumar, G.B. and P.N. Black. 1993. Bacterial long-chain fatty acid transport: Identification of amino acid residues within the outer membrane protein FadL required for activity. *J. Biol. Chem.,* 268:15469–15476.

Lam, S.T., D.M. Ellis, and J. Ligon. 1991. Genetic approaches for studying rhizosphere colonization. In *The Rhizosphere and Plant Growth*. D.L. Keister and P.B. Cregan, Eds. Kluwer Academic Publishers, Amsterdam.

Laville, J., C. Voisard, C. Keel, M. Maurhofer, G. Défago, and D. Haas. 1992. Global control in *Pseudomonas fluorescens* mediating antibiotic synthesis and suppression of black root rot of tobacco. *Proc. Natl. Acad. Sci. USA,* 89:1562–1566.

Lawton, M.B. and L.L. Burpee. 1990. Effect of rate and frequency of application of *Typhula phacorrhiza* on biological control of *Typhula* blight of creeping bentgrass. *Phytopathology,* 80:70–73.

Lawton, M.B., L.L. Burpee, and L.G. Goulty. 1987. Influence of several factors on the biocontrol of gray snow mold of turfgrass by *Typhula phacorrhiza. Phytopathology,* 77:119.

Leeman, M., F.M. den Ouden, J.A. van Pelt, F.P.M. Dirkx, H. Steijl, P.A.H.M. Bakker, and B. Schippers. 1996. Iron availability affects induction of systemic resistance to Fusarium wilt of radish by *Pseudomonas fluorescens. Phytopathology,* 86:149–155.

Lemanceau, P., P.A.H.M. Bakker, W.J. Dekogel, C. Alabouvette, and B. Schippers. 1992. Effect of pseudobactin-358 production by *Pseudomonas putida* WCS358 on suppression of Fusarium wilt of carnations by nonpathogenic *Fusarium oxysporum* Fo47. *Appl. Environ. Microbiol.,* 58:2978–2982.

Leong, J. 1986. Siderophores: their biochemistry and possible role in the biological control of plant pathogens. *Ann. Rev. Phytopathol.,* 24:187–209.

Lifshitz, R., M.T. Windham, and R. Baker. 1986. Mechanism of biological control of preemergence damping-off of pea by seed treatment with *Trichoderma* spp. *Phytopathology,* 76:720–725.

Lo, C.-T., E.B. Nelson, and G.E. Harman. Unpublished.

Lo, C.-T., E.B. Nelson, and G.E. Harman. 1996. Biological control of turfgrass diseases with a rhizosphere competent strain of *Trichoderma harzianum. Plant Dis.,* 80:736–741.

Lo, C.-T., E.B. Nelson, and G.E. Harman. 1997. Improving the biocontrol efficacy of *Trichoderma harzianum* for controlling foliar phases of turf diseases by spray applications. *Plant Dis.,* in press.

Loper, J.E. 1988. Role of fluorescent siderophore production in biological control of *Pythium ultimum* by a *Pseudomonas fluorescens* strain. *Phytopathology,* 78:166–172.

Loper, J.E. and J.S. Buyer. 1991. Siderophores in microbial interactions on plant surfaces. *Molec. Plant-Microbe Interact.,* 4:5–13.

Loper, J.E. and S.E. Lindow. 1994. A biological sensor for iron available to bacteria in their habitats on plant surfaces. *Appl. Environ. Microbiol.,* 60:1934–1941.

Lorito, M., G.E. Harman, C.K. Hayes, R.M. Broadway, A. Tronsmo, S.L. Woo, and A. Di Pietro. 1993a. Chitinolytic enzymes produced by *Trichoderma harzianum*: Antifungal activity of purified endochitinase and chitobiosidase. *Phytopathology,* 83:302–307.

Lorito, M., C.K. Hayes, A. Di Pietro, and G.E. Harman. 1993b. Biolistic transformation of *Trichoderma harzianum and Gliocladium virens* using plasmid and genomic DNA. *Current Genetics,* 24:349–356.

Lucas, P. and A. Sarniguet. Screening in the greenhouse of some treatments to control take-all patch on turfgrass. *Phytopathology,* 81:1198, 1991.

Lucas, P., A. Sarniguet, N. Cavelier, and S. Lelarge. 1992a. Etude preliminaire sur l'efficacité de differents moyens de lutte contre le pietin-échaudage du gazon (*Gaeumannomyces graminis* var. *avenae*). *Agronomie,* 12:187–192.

Lucas, P., A. Sarniguet, and C. Laurent. 1992b. Manifestation, en France, du pietin-échaudage sur gazon du a *Gaeumannomyces graminis* var. *avenae*. *Agronomie,* 12:183–186.

Lucas, P., A. Sarniguet, and S. Lelarge. Bacterial populations related to the progress of the disease in take-all patches on turf grass. In *Biotic Interactions and Soilborne Diseases.* A.B.M. Beemster, G.J. Bollen, M. Gerlagh, M.A. Ruissen, B. Schippers, and A. Tempel, Eds. Elsevier Publishing Company, Amsterdam, 1991.

Lynch, J.M., R.D. Lumsden, P.T. Atkey, and M.A. Ousley. 1991. Prospects for control of *Pythium* damping-off of lettuce with *Trichoderma, Gliocladium,* and *Enterobacter* spp. *Biol. Fertil. Soils,* 12:95–99.

Maloney A.P. and E.B. Nelson. Unpublished.

Maloney, A.P. and E.B. Nelson. 1992. Relationship between the biological control of Pythium seed and seedling disease and adherence to fungal structures by the bacterium *Enterobacter cloacae*. *Phytopathology,* 82:1119.

Maloney, A.P. and E.B. Nelson. 1994. Isolation of a gene from *Enterobacter cloacae* that affects biological control of *Pythium ultimum* seed rot. *Phytopathology,* 84:1082–1083.

Maloney, A.P., E.B. Nelson, and K. van Dijk. 1994. Genetic complementation of a biocontrol-negative mutant of *Enterobacter cloacae* reveals a potential role of pathogen stimulant inactivation in the biological control of Pythium seed rots. In *Improving Plant Productivity with Rhizosphere Bacteria.* M.H. Ryder, P.M. Stephens, and G.D. Bowen, Eds. Graphic Services, Adelaide.

Mancino, C.F., M. Barakat, and A. Maricic. 1993. Soil and thatch microbial populations in an 80% sand: 20% peat creeping bentgrass putting green. *Hortscience,* 28:189–191.

Manoil, C. and J. Beckwith. 1985. Tn*phoA*: a transposon probe for protein export signals. *Proc. Natl. Acad. Sci. USA,* 82:8129–8133.

Manoil, C., J.J. Mekalanos, and J. Beckwith. 1990. Alkaline phosphatase fusions: sensors of subcellular location. *J. Bacteriol.,* 172:515–518.

Maurhofer, M., C. Keel, D. Haas, and G. Defago. 1994. Pyoluteorin production by *Pseudomonas fluorescens* strain CHA0 is involved in the suppression of Pythium damping-off of cress but not of cucumber. *Eur. J. Plant Pathol.,* 100:221–232.

Mazzola, M., R.J. Cook, L.S. Thomashow, D.M. Weller, and L.S. Pierson. 1992. Contribution of phenazine antibiotic biosynthesis to the ecological competence of fluorescent Pseudomonads in soil habitats. *Appl. Environ. Microbiol.,* 58:2616–2624.

Meyer, J.M., P. Azelvandre, and C. Georges. 1992. Iron metabolism in *Pseudomonas* — salicylic acid, a siderophore of *Pseudomonas fluorescens* CHA0. *Biofactors,* 4:23–27.

Nelson, E.B. 1987. Rapid germination of sporangia of *Pythium* species in response to volatiles from germinating seeds. *Phytopathology,* 77:1108–1112.

Nelson, E.B. 1988. Biological control of Pythium seed rot and preemergence damping-off of cotton with *Enterobacter cloacae* and *Erwinia herbicola* applied as seed treatments. *Plant Dis.,* 72:140–142.

Nelson, E.B. 1990. Exudate molecules initiating fungal responses to seeds and roots. *Plant Soil,* 129:61–73.

Nelson, E.B. 1992. Bacterial metabolism of propagule germination stimulants as an important trait in the biocontrol of Pythium seed infections. In *Biological Control of Plant Diseases: Progress and Challenges for the Future.* E.C. Tjamos, G.C. Papavizas, and R.J. Cook, Eds. Plenum Publishing Company, New York.

Nelson, E.B. 1996. Enhancing turfgrass disease control with organic amendments. *Turfgrass TRENDS,* 5(6):1–15.

Nelson, E.B., L.L. Burpee, and M.B. Lawton. 1994. Biological control of turfgrass diseases. In *Integrated Pest Management for Turfgrass (and Ornamentals).* A. Leslie, Ed. Lewis Publishers, Boca Raton, FL.

Nelson, E.B., W.L. Chao, J.M. Norton, G.T. Nash, and G.E. Harman. 1986. Attachment of *Enterobacter cloacae* to hyphae of *Pythium ultimum*: Possible role in the biological control of Pythium preemergence damping-off. *Phytopathology,* 76:327–335.

Nelson, E.B. and C.M. Craft. 1991. Introduction and establishment of strains of *Enterobacter cloacae* in golf course turf for the biological control of dollar spot. *Plant Dis.,* 75:510–514.

Nelson, E.B. and C.M. Craft. 1992. A miniaturized and rapid bioassay for the selection of soil bacteria suppressive to Pythium blight of turfgrasses. *Phytopathology,* 82:206–210.

Nelson, E.B. and A.P. Maloney. Molecular approaches for understanding biological control mechanisms in bacteria: Studies of the interaction of *Enterobacter cloacae* with *Pythium ultimum. Can. J. Plant Pathol.,* 14:106–114, 1992.

Nelson, E.B. and T.R. Ruttledge. Unpublished.

Nowak-Thompson, B., S.J. Gould, J. Kraus, and J.E. Loper. 1994. Production of 2,4-diacetylphloroglucinol by the biocontrol agent *Pseudomonas fluorescens* Pf-5. *Can. J. Microbiol.,* 40:1064–1066.

Nunn, W.D. 1986. A molecular view of fatty acid catabolism in *Escherichia coli. Microbiol. Rev.,* 50:179–192.

Nunn, W.D. 1987. Utilization of 2-carbon compounds: fatty acids. In *Escherichia coli and Salmonella typhimurium: Cellular and Molecular Biology.* J.L.

Ingraham, K.B. Low, B. Magasanik, M. Schaechter, and H.E. Umbarger, Eds. American Society for Microbiology, Washington.

O'Leary, A.L., D.J. O'Leary, and S.H. Woodhead. 1988. Screening potential bioantagonists against pathogens of turf. *Phytopathology,* 78:1593.

Osburn, R.M. and M.N. Schroth. 1988. Effect of osmopriming sugar beet seed on exudation and subsequent damping-off caused by *Pythium ultimum. Phytopathology,* 78:1246–1250.

Osburn, R.M., M.N. Schroth, J.G. Hancock, and M. Hendson. 1989. Dynamics of sugar beet seed colonization by *Pythium ultimum* and *Pseudomonas* species: effects on seed rot and damping-off. *Phytopathology,* 79:709–716.

Ossana, N. and S. Mischke. 1990. Genetic transformation of the biocontrol fungus *Gliocladium virens* to benomyl resistance. *Appl. Environ. Microbiol.,* 56:3052–3056.

Ownley, B.H., D.M. Weller, and L.S. Thomashow. 1992. Influence of in situ and in vitro pH on suppression of *Gaeumannomyces graminis* var *tritici* by *Pseudomonas fluorescens* 2-79. *Phytopathology,* 82:178–184.

Paulitz, T.C. 1991. Effect of *Pseudomonas putida* on the stimulation of *Pythium ultimum* by seed volatiles of pea and soybean. *Phytopathology,* 81:1282–1287.

Paulitz, T.C., O. Anas, and D.G. Fernando. 1992. Biological control of Pythium damping-off by seed-treatment with *Pseudomonas putida* — relationship with ethanol production by pea and soybean seeds. *Biocont. Sci. Technol.,* 2:193–201.

Paulitz, T.C. and J.E. Loper. 1991. Lack of a role for fluorescent siderophore production in the biological control of cucumber by a strain of *Pseudomonas putida. Phytopathology,* 81:930–935.

Pe'er, S. and I. Chet. 1990. *Trichoderma* protoplast fusion: a tool for improving biocontrol agents. *Can. J. Microbiol.,* 36:6–9.

Peacock, C.H. and P.F. Daniel. 1992. A comparison of turfgrass response to biologically amended fertilizers. *Hortscience,* 27:883–884.

Pentilla, M., H. Nevalainen, M. Rättö, E. Salminen, and J. Knowles. 1987. A versatile transformation system for the cellulolytic filamentous fungus *Trichoderma reesei. Gene,* 61:155–164,

Pierson, L.S., V.D. Keppenne, and D.W. Wood. 1994. Phenazine antibiotic biosynthesis in *Pseudomonas aureofaciens* 30-84 is regulated by PhzR in response to cell density. *J. Bacteriol.,* 176:3966–3974.

Pierson, L.S. and L.S. Thomashow. 1991. Analysis of phenazine antibiotic production by *Pseudomonas aureofaciens* strain 30-84. In *Plant Growth-Promoting Rhizobacteria: Progress and Prospects.* C. Keel, B. Koller, and G. Défago, Eds.

Punja, Z.K., R.G. Grogan, and T. Unruh. 1982. Comparative control of *Sclerotium rolfsii* on golf greens in North Carolina with fungicides, inorganic salts, and *Trichoderma* spp. *Plant Dis.,* 66:1125–1128.

Rasmussen-Dykes, C. and W.M. Brown, Jr. 1982. Integrated control of Pythium blight on turf using metalaxyl and *Trichoderma hamatum. Phytopathology,* 72:974.

Reuter, H.M., G.L. Schumann, M.L. Matheny, and R.T. Hatch. 1991. Suppression of dollar spot (*Sclerotinia homoeocarpa*) and brown patch (*Rhizoctonia solani*) on creeping bentgrass by an isolate of *Streptomyces*. *Phytopathology,* 81:124.

Roberts, D.P., C.J. Sheets, and J.S. Hartung. 1992. Evidence for proliferation of *Enterobacter cloacae* on carbohydrates in cucumber and pea spermosphere. *Can. J. Microbiol.,* 38:1128–1134.

Roberts, D.P., N.M. Short, Jr., A.P. Maloney, E.B. Nelson, and D.A. Schaff. 1994. Role of colonization in biocontrol: studies with *Enterobacter cloacae*. *Plant Sci.,* 101:83–89.

Ross, I.L. and M.H. Ryder. 1994. Hydrogen cyanide production by a biocontrol strain of *Pseudomonas corrugata*: evidence that cyanide antagonises the take-all fungus *in vitro*. In *Improving Plant Productivity with Rhizosphere Bacteria*. M.H. Ryder, P.M. Stephens, and G.D. Bowen, Eds. Graphic Services, Adelaide.

Ruttledge, T.R. and E.B. Nelson. Unpublished.

Ruttledge, T.R. and E.B. Nelson. 1997. Exudate Fatty Acids from *Gossypium hirsutum* Stimulatory to the Seed-Rotting Fungus, *Pythium ultimum*. *Phytochem,* in press.

Ruvkin, G.B. and F.M. Ausubel. 1981. A general method for site-directed mutagenesis in prokaryotes. *Nature,* 289:85–88.

Sanders, P.L. and M.D. Soika. 1991. Biological control of Pythium blight, 1990. *Biol. Cult. Tests Cont. Plant Dis.,* 6:99.

Sarniguet, A. and J.E. Loper. 1994. A *rpoS*-like sigma factor gene is involved in pyrollnitrin production by *Pseudomonas fluorescens* Pf-5. *Phytopathology,* 84:1134.

Sarniguet, A. and P. Lucas. 1991. Evolution of bacterial populations related to decline of take-all patch on turfgrass. *Phytopathology,* 81:1202.

Schumann, G.L. and H.M. Reuter. 1993. Suppression of dollar spot with wheat bran topdressings. *Biol. Cult. Tests Cont. Plant Dis.,* 7:113.

Segall, L. 1995. Marketing compost as a pest control product. *BioCycle,* May 1995, 65–67.

Shapira, R., A. Ordentlich, I. Chet, and A.B. Oppenheim. 1989. Control of plant diseases by chitinase expressed from cloned DNA in *Escherichia coli*. *Phytopathology,* 79:1246–1249.

Shaw, J.J. and C.I. Kado. 1986. Development of a *Vibrio* bioluminescence gene set to monitor phytopathogenic bacteria during the ongoing disease process in a nondestructive manner. *Bio/Technology,* 4:560–564.

Silhavy, T.J., M.L. Berman, and L.W. Enquist. 1984. Experiments with gene fusions. Cold Spring Harbor Laboratory. Cold Spring Harbor, NY.

Sivan, A. and G.E. Harman. 1991. Improved rhizosphere competence in a protoplast fusion progeny of *Trichoderma harzianum*. *J. Gen. Microbiol.,* 137:23–29.

Sivan, A., T.E. Stasz, M. Hemmat, C.K. Hayes, and G.E. Harman. 1992. Transformation of *Trichoderma* spp. with plasmids conferring hygromycin B resistance. *Mycologia,* 84:687–694.

Slininger, P.J. and M.A. Jackson. 1992. Nutritional factors regulating growth and accumulation of phenazine 1-carboxylic acid by *Pseudomonas fluorescens* 2-79. *Appl. Microbiol. Biotechnol.*, 37:388–392.

Slininger, P.J. and M.A. Sheawilbur. 1995. Liquid culture pH, temperature, and carbon (not nitrogen) source regulate phenazine productivity of the take-all biocontrol agent *Pseudomonas fluorescens* 2-79. *Appl. Microbiol. Biotechnol.*, 43:794–800.

Sneh, B., M. Dupler, Y. Elad, and R. Baker. 1984. Chlamydospore germination of *Fusarium oxysporum* f.sp. *cucumerinum* as affected by fluorescent and lytic bacteria from a Fusarium-suppressive soil. *Phytopathology*, 74:1115–1124.

Soika, M.D. and P.L. Sanders. 1991. Effect of various nitrogen sources, organic amendments, and biological control agents on turfgrass quality and disease development, 1990. *Biol. Cult. Tests Cont. Plant Dis.*, 6:91.

Soika, M.D. and P.L. Sanders. 1992. Biological control of Pythium blight, 1991. *Biol. Cult. Tests Cont. Plant Dis.*, 7:113.

Stanghellini, M.E. and J.G. Hancock. 1971. Radial extent of the bean spermosphere and its relation to the behavior of *Pythium ultimum*. *Phytopathology*, 61:165–168.

Stasz, T.E. and G.E. Harman. 1990. Nonparental progeny resulting from protoplast fusion in *Trichoderma* in the absence of parasexuality. *Exp. Mycol.*, 14:145–159.

Stasz, T.E., G.E. Harman, and N.F. Weeden. 1988. Protoplast preparation and fusion in two biocontrol strains of *Trichoderma harzianum*. *Mycologia*, 80:141–150.

Stockwell, C.T., E.B. Nelson, and C.M. Craft. 1994. Biological control of *Pythium graminicola* and other soilborne pathogens of turfgrass with actinomycetes from composts. *Phytopathology*, 84:1113.

Sundheim, L., A.R. Poplawsky, and A.H. Ellingboe. 1988. Molecular cloning of two chitinase genes from *Serratia marcescens* and their expression in *Pseudomonas* species. *Physiol. Molec. Plant Pathol.*, 33:483–491.

Sutker, E.M. and L.T. Lucas. 1987. Biocontrol of *Rhizoctonia solani* in tall fescue turfgrass. *Phytopathology*, 77:1721.

Taylor, A.G., Y. Hadar, J.M. Norton, A.A. Khan, and G.E. Harman. 1985. Influence of presowing seed treatments of table beets on the susceptibility to damping-off caused by *Pythium*. *J. Am. Soc. Hort. Sci.*, 110:516–519.

Thomas, M. and C. Kennerley. 1989. Transformation of the mycoparasite *Gliocladium*. *Current Genetics*, 15:415–420.

Thomashow, L.S. 1990. Production of the antibiotic phenazine-1-carboxylic acid by fluorescent *Pseudomonas* species in the rhizosphere of wheat. *Appl. Environ. Microbiol.*, 56:908–912.

Thomashow, L.S. and L.S. Pierson. 1991. Genetic aspects of phenazine antibiotic production by fluorescent pseudomonads that suppress take-all disease of wheat. In *Advances in Molecular Genetics of Plant-Microbe Interactions, Vol 1*. H. Hennecke and D.P.S. Verma, Eds. Kluwer Academic Publishers, Dordrecht, Netherlands.

Thomashow, L.S. and D.M. Weller. 1988. Role of a phenazine antibiotic from *Pseudomonas fluorescens* in biological control of *Gaeumannomyces graminis* var. *tritici*. *J. Bacteriol.,* 170:3499–3508.

Thompson, D.C. and B.B. Clarke. 1992. Evaluation of bacteria for biological control of summer patch of Kentucky bluegrass caused by *Magneporthe poae*. *Phytopathology,* 82:1123.

Thompson, D.C., D.Y. Kobayashi, and B.B. Clarke. 1993. Evaluation of bacteria for the suppression of summer patch and root colonizing ability on turfgrass. *Phytopathology,* 83:1337.

Trutmann, P. and E.B. Nelson. 1992. Production of non-volatile and volatile inhibitors of *Pythium ultimum* sporangium germination and mycelial growth by strains of *Enterobacter cloacae*. *Phytopathology,* 82:1120.

Tucker, W.T., S.J. Abbene, and N. Gutterson. 1988. Isolation of genes for the biosynthesis of fusaromycin A, an antibiotic active against *Fusarium* and *Thielaviopsis*. *Phytopathology,* 78:1587.

Tuzun, S. and J. Kloepper. 1994. Induced systemic resistance by plant growth-promoting rhizobacteria. In *Improving Plant Productivity with Rhizosphere Bacteria*. M.H. Ryder, P.M. Stephens, and G.D. Bowen, Eds. Graphic Services, Adelaide.

van Dijk, K. 1995. Seed exudate stimulant inactivation by *Enterobacter cloacae* and its involvement in the biological control of *Pythium ultimum.* MS Thesis, Cornell University, 96 pp.

Villani, M.G. 1995. What's new in turfgrass insect pest management products: Focus on biological controls. *Turfgrass TRENDS,* 4(6):1–6.

Voisard, C., C. Keel, D. Haas, and G. Défago. 1989. Cyanide production by *Pseudomonas fluorescens* helps suppress black root rot of tobacco under gnotobiotic conditions. *EMBO J.,* 8:351–358.

Wilhite, S.E., R.D. Lumsden, and D.C. Straney. 1994. Mutational analysis of gliotoxin production by the biocontrol fungus *Gliocladium virens* in relation to suppression of Pythium damping-off. *Phytopathology,* 84:816–821.

Wilkinson, H.T. and R. Avenius. 1985. The selection of bacteria antagonistic to *Pythium* spp. pathogenic to turfgrass. *Phytopathology,* 75:812.

Wilson, C.L., J.D. Franklin, and P.L. Pusey. 1987. Biological control of Rhizopus soft rot of peach with *Enterobacter cloacae*. *Phytopathology,* 77:303–305.

Wisniewski, M., C. Wilson, and W. Hershberger. 1989. Characterization of inhibition of *Rhizopus stolonifer* germination and growth by *Enterobacter cloacae*. *Can. J. Bot.,* 67:2317–2323.

Wong, P.T.W. and R. Baker. 1984. Suppression of wheat take-all and Ophiobolus patch by fluorescent pseudomonads from a Fusarium-suppressive soil. *Soil Biol. Biochem.,* 16:397–403.

Wong, P.T.W. and R. Baker. 1985. Control of wheat take-all and Ophiobolus patch of turfgrass by fluorescent pseudomonads. In *Ecology and Management of Soilborne Plant Pathogens*. C.A. Parker, A.D. Rovira, K.J. Moore, P.T.W. Wong, and J.F. Kollmorgen, Eds. The American Phytopathological Society, St. Paul, Minnesota.

Wong, P.T.W. and T.R. Siviour. 1979. Control of Ophiobolus patch in *Agrostis* turf using avirulent fungi and take-all suppressive soils in pot experiments. *Ann. Appl. Biol.,* 92:191–197.

Wong, P.T.W. and D.J. Worrand. 1989. Preventative control of take-all patch of bentgrass turf using triazole fungicides and *Gaeumannomyces graminis* var. *graminis* following soil fumigation. *Plant Protection Quarterly,* 4:70–72.

Xi, K., J.H.G. Stephens, and S.F. Hwang. 1995. Dynamics of pea seed infection by *Pythium ultimum* and *Rhizoctonia solani*: Effects of inoculum density and temperature on seed rot and pre-emergence damping-off. *Can. J. Plant Pathol.,* 17:19–24.

Yuen, G.Y. and M.L. Craig. 1992. Reduction in brown patch severity by a binucleate *Rhizoctonia* antagonist, 1991. *Biol. Cult. Tests Cont. Plant Dis.,* 7:114.

Yuen, G.Y., M.L. Craig, and L.J. Giesler. 1994. Biological control of *Rhizoctonia solani* on tall fescue using fungal antagonists. *Plant Dis.,* 78:118–123.

Zhou, T. and J.C. Neal. 1995. Annual bluegrass (*Poa annua*) control with *Xanthomonas campestris* pv *poannua* in New York state. *Weed Technol.,* 9:173–177.

Zhu, Q., E.A. Maher, S. Masoud, R.A. Dixon, and C.J. Lamb. 1994. Enhanced protection against fungal attack by constitutive co-expression of chitinase and glucanase genes in transgenic tobacco. *Bio-Technology,* 12:807–812.

Chapter 6

Biological Control of Turfgrass Diseases: Past and Future

J.M. Vargas, Jr., J.F. Powell, N.M. Dykema, A.R. Detweiler, and M.G. Nair

INTRODUCTION

Biological control in the past has mainly consisted of the use of composted materials and organic fertilizers. The results have been mixed, at best. Often, when controlled-release fertilizers were compared to composted products and organic fertilizers, similar disease control was achieved. In most cases, with diseases that occur on golf course turfs, the level of control obtained with the composted materials was not satisfactory and did not eliminate fungicide treatments during years of moderate to severe disease pressure.

More recently, attempts have been made to add specific antagonists to the turf, either alone or with composted materials or organic fertilizers. Lo et al. (1996), in an article describing work they had done using *Trichoderma harzianum,* have shown suppression of root diseases, like Pythium root rot, with early spring applications. However, they had less success in controlling foliar diseases like dollar spot.

Similar results were experienced when *Psuedomonas aureofaciens* strains Tx-1 and Tx-2 were applied alone or in combination with organic products on a biweekly basis (Powell, 1993). It appears that, for control of foliar turfgrass pathogens, the frequency of application will have to be increased. This is due to the inability of the of the antagonist to become established in the soil.

Bacterial strain Tx-1 was initially isolated at Michigan State University and proved to exhibit antagonism toward a broad spectrum of fungal turfgrass pathogens in laboratory plate bioassays. Based on these results, this organism was chosen for further evaluation as a potential biological control agent for the management of turfgrass diseases. This study examines the efficacy of Tx-1 as a biological control agent for *Sclerotinia homoeocarpa*, the causal agent of dollar spot.

MATERIALS AND METHODS

Two methods of preparing the bacteria were employed in this study. One involved daily culturing of the bacteria 24 hours prior to application. This application procedure would assure that the bacteria would be metabolically active upon application. The other method involved the culturing of the bacteria prior to the onset of the field season. These bacteria were stored in refrigeration until the time of application. One subset of this group was amended with formulating agents to aid in the stability of the organism upon application.

Field evaluation of Tx-1 for the management of dollar spot was conducted at the Hancock Turfgrass Research Center in East Lansing, Michigan, on a stand of Penncross creeping bentgrass naturally infested with *S. homoeocarpa*. The study was initiated July 30, 1995, and plots were maintained at 0.5 lb nitrogen/1000 ft^2 per month (fertility was not applied during the month of August.) Bacterial treatments included in this study were applied three times a week using a CO_2 hand sprayer. The study was established in a randomized complete block design with four replications and plots that were 0.9 m $\times$ 1.2 m. Data were recorded on a weekly basis as the number of spots within a plot.

The bacteria used in this study were cultured by two methods. Daily cultures of bacteria were grown overnight in 2 L trypticase soy broth in a 4 L Erlenmeyer flask at room temperature, under constant aeration. Bacterial cultures used to examine formulating agents were prepared prior to initiation of the field study. These cultures were grown in A9 media (20 g Brer Rabbit molasses, 10 g dextrose, and 5 g peptone) in a 10 L Braun fermenter at 26°C for a period of 72 hr. Following incubation, the bacterial cultures were divided into unformulated and formulated treatments. Formulated treatments were amended with chemicals to aid in culture stability when applied in the field. Formulated and unformulated cultures were stored at 4°C until time of application.

Quantitation of bacteria for field application was conducted by pelleting cells from 50 mL of broth by centrifugation. The resulting pellet was resuspended in tap water and the optical density was measured using a Spec 20 spectrophotometer at 640 nm. An appropriate amount of culture was applied to deliver 2×10^5 or 2×10^7 cfu per cm^2.

Control treatments consisted of autoclaved cultures of fresh or refrigerated bacteria to account for possible fertility effects by the bacteria or accompanying media. Autoclaved cultures were applied to deliver the same volumes as bacterial treatments at the 2×10^7 cfu/cm^2 rate. A chemical control was also included which consisted of a spray application of chlorothalonil (Daconil 2787 WDG) at a rate of 3 oz/1000 ft^2 every 10 days.

Variation in collected data was normalized as the log (number of spots + 1). Data was analyzed by Tukey's Honestly Significant range test using MSTAT-C statistical software.

RESULTS AND DISCUSSION

Dollar spot disease pressure was not significant until the September 6 rating date and persisted through to early October. Dollar spot ratings are provided in Table 6.1. Application of the precultured bacteria failed to provide significant disease reduction. This may be due to a loss of viability during storage and application, or it may also be attributed to the bacteria not being metabolically competitive following storage under refrigeration. Only the chemical control, chlorothalonil, and application of the freshly grown bacteria at a rate of 2×10^7 cfu/cm^2 provided significant reduction in disease incidence with respect to the control. The fresh bacteria at the 2×10^7 cfu/cm^2 rate was also significantly better than the control treatment consisting of autoclaved freshly grown bacteria, indicating that disease reduction was not due to fertility effects.

This study demonstrates that the bacterial strain Tx-1 is efficacious as a biological control agent for the management of dollar spot when applied frequently at 2×10^7 cfu/cm^2. Such frequent applications are feasible with the advent of the BioJect system (EcoSoil Systems, Inc., San Diego, California) which couples a bacterial fermentation system to an established irrigation system. One handicap of this study is that the bacteria were often applied during mid-day, during which time the bacteria are most susceptible to UV light. Bacterial applications made through an irrigation system should be made at night when the dollar spot fungus is active and there is no UV light. The use of Tx-1 along with higher fertility rates (1 lb N/month) may well provide dollar spot management at levels provided by chemical fungicides and should be investigated in future studies.

The future of biological control should involve the genetic engineering of biological control agents that are more efficient producers of antifungal compounds and organisms that are more fit to survive in the environment. This may help overcome the greatest drawback to successful biological control — establishing the antagonist in the soil.

REFERENCES

1. Lo, C.-T., E.B. Nelson, and G.E. Harman. 1996. Biological Control of Turfgrass Diseases with a Rhizosphere Competent Strain of *Trichoderma harzianum*. *Plant Disease,* 80:736–741.
2. Powell, J.P. 1993. Utilization of Bacterial Metabolites for the Management of Fungal Turfgrass Pathogens (MS Thesis). Michigan State University.

Table 6.1. Suppression of Dollar Spot with Fresh and Refrigerated Cultures of *Pseudomonas aureofaciens*.

		September 30, 1995			October 4, 1995		
Treatment	**Rate[a]**	**Average No. of Spots**	**Normalized[b] Data Analysis**		**Average No. of Spots**	**Normalized Data Analysis**	
No Treatment	—	24.3	1.40	A	32.5	1.52	A
Fresh Bacteria[c]	2×10^5 cfu/cm^2	20.0	1.27	AB	21.8	1.31	AB
Fresh Bacteria	2×10^7 cfu/cm^2	11.0	1.04	B	13.5	1.09	B
Autoclaved Broth[d]	Volume as Above	24.5	1.41	A	34.8	1.55	A
Refrigerated Bacteria[e]	2×10^5 cfu/cm^2 (Unformulated)	20.5	1.32	AB	19.5	1.30	AB
Refrigerated Bacteria	2×10^7 cfu/cm^2 (Formulated)	24.8	1.39	A	27.0	1.43	AB
Refrigerated Bacteria	2×10^5 cfu/cm^2 (Unformulated)	22.8	1.36	A	24.8	1.41	AB
Refrigerated Bacteria	2×10^7 cfu/cm^2 (Formulated)	24.0	1.36	A	28.8	1.42	AB
Sterile Broth[f]	Volume as Above	19.8	1.30	A	21.8	1.36	AB
Chlorothalonil	0.15 g AI/m^2	0.3	0.08	C	0.8	0.19	C
	Standard Deviation	2.8	0.06		4.0	0.08	

[a] Application rates made on plots 0.9 m × 1.2 m.
[b] Data transformation performed as log(# of spots + 1).
[c] Bacteria cultured on TSB 24 hr prior to application.
[d] Bacterial broth autoclaved after 24 hr growth and applied at a volume equal to 2×10^7 cfu/cm^2 rate.
[e] Bacteria cultured on A9 broth for 72 hr and stored at 4°C until time of application.
[f] Sterile A9 broth applied at a volume equal to 2×10^7 cfu/cm^2 (unformulated) rate.

Chapter 7

The Use of Endophytes to Improve Turfgrass Performance

M.D. Richardson, J.F. White, Jr., and F.C. Belanger

CLASSIFICATION AND BIOLOGY OF ENDOPHYTES

Most grass endophytes belong in the Ascomycete family Clavicipitaceae (Ascomycetes). Endophytes are widespread in the grasses (White, 1988; Clay and Leuchtmann, 1989) and several genera of endophytes are commonly encountered. While endophytes of the genus *Balansia* are common in warm-season grasses, these endophytes are not transmitted through the seed and have limited applications in turf. Endophytes of the genera *Epichloë* and *Neotyphodium* (=*Acremonium*) are found in many cool-season turfgrasses, are frequently seed-transmitted, and have more immediate applications to turfgrass development than *Balansia* endophytes.

Clavicipitaceous endophytes perennate as unbranching, often convoluted, mycelium in the intercellular spaces of grass plants (Figure 7.1). Endophytic mycelium is distributed in the leaf sheaths, reproductive culms, and seeds. Many endophytes, such as those found in tall fescue and perennial ryegrass, are strictly seed-transmitted and are dependent on host seed production for dissemination. As the host plant is changed from vegetative growth to the reproductive phase, the endophyte moves with the apical meristem into the developing inflorescence and subsequently into the seed. Once in the seed, the endophyte proliferates in the aleurone layer of the seed, where it will persist until the seed germinates and the fungus reinfects the new seedling. Although completely heterotrophic throughout its life cycle, these seed-transmitted en-

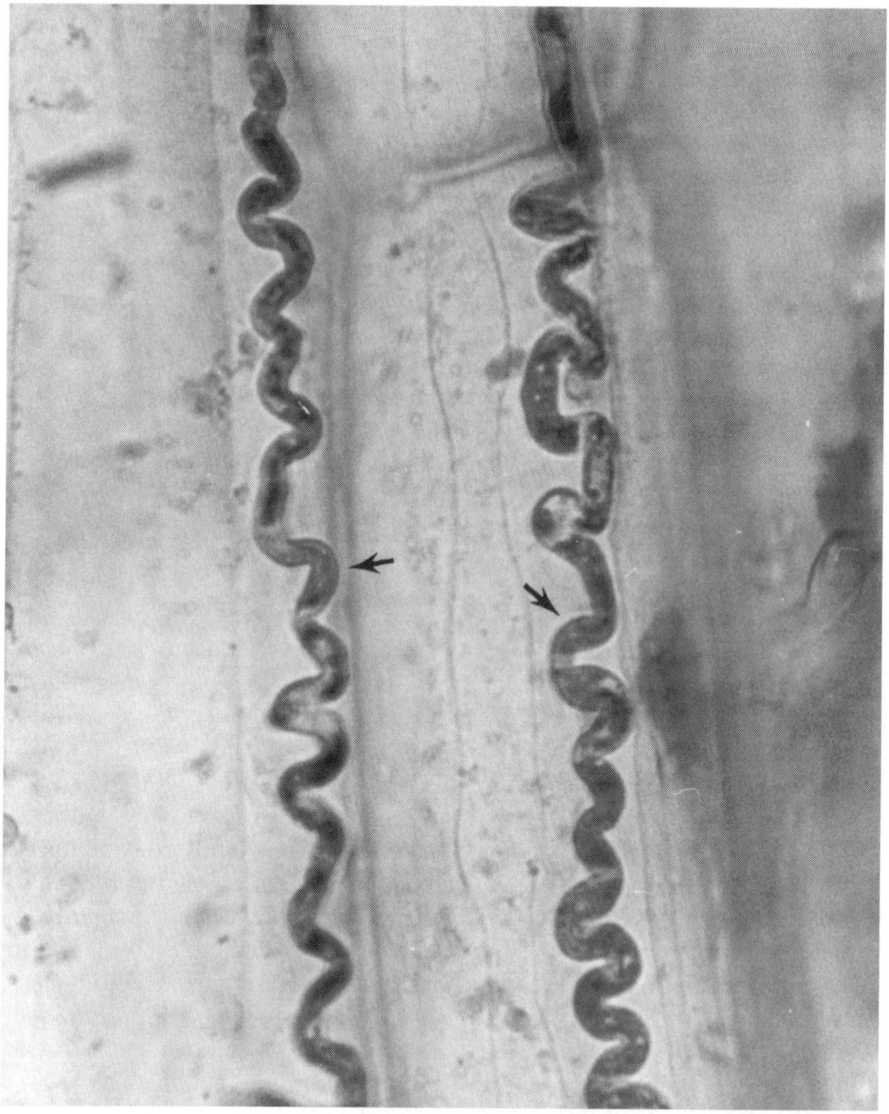

Figure 7.1. Mycelium (arrows) of endophyte *Neotyphodium coenophialum* in culm of tall fescue grass (X 3,000).

dophytes apparently require very little energy from their host (Richardson et al., 1993) and cause no apparent pathological symptoms or stresses due to parasitism.

In other endophytes, an aggressive, epiphytic stage called a stroma or "choke" is produced. The stroma is the primary reproductive structure of *Epichloë* and all of the clavicipitaceous endophytes. This is the structure on which endophytes develop their reproductive cells (spermatia and perithecia) and, in the process, extract large quantities of energy from the host plant to fuel development. Culms of endophyte-infected *Agrostis hiemalis* bearing stromata were found to have on average 44% less mass than endophyte-infected culms without stromata (White and Camp, 1995). It is reasonable to assume that a good part of this reduction in mass is the result of the diversion of substrates from the process of plant inflorescence development to the fungus for use in stroma construction. As a result of this energy utilization and abortion of the host reproductive cycle, these fungi must be considered pathogenic during the stroma phase.

THE OCCURRENCE OF TOXICITIES

Endophytic fungi of grasses have been studied for up to 100 years. However, their importance has escalated upon the discovery that these fungi are responsible for widespread livestock poisonings throughout the world (Bacon et al., 1977; Fletcher and Harvey, 1981). In the U.S. beef market alone, annual economic losses associated with these fungi have been estimated to be in excess of $600 million (Hoveland, 1992). Endophyte-infected grasses can be toxic to livestock because the fungus produces a wide range of chemicals (Table 7.1), many of which have a high degree of biological activity against mammalian systems. The most thoroughly studied compounds are alkaloids, including ergopeptine alkaloids, indole-isoprenoid lolitrems, pyrrolizidine alkaloids, and pyrrolopyrazine alkaloids (Porter et al., 1981; Yates et al., 1990; Rowan and Gaynor, 1986; Siegel et al., 1990).

The ergopeptine alkaloids were once thought to be only associated with the ergot fungus, *Claviceps*, but they are now known to exist in other fungi, such as *Aspergillus* and *Penicillium,* and have also been found in a few isolated higher plants, primarily the *Convolvulaceae* (morning glory family; Rehacek and Sajdl, 1990). Lyons et al. (1986) were the first to demonstrate the occurrence of ergopeptine alkaloids in endophyte-infected grasses, and these alkaloids are now believed to be the major toxins associated with many symbiotic grasses (Porter, 1994). Ergopeptines have been identified in a range of endophyte-infected grasses, including *Festuca arundinacea* Schreb., *Lolium perenne* L., *F. longifolia* Thuill., *F. glauca* Vill., *F. rubra*, *Bromus anomalus* Rupt., and *Agrostis hiemalis* (Walt.) B.S.P. (Siegel et al., 1990; Yates and Powell, 1988). Although the ergopeptine alkaloids have a high degree of biological activity against animals (Rehacek and Sajdl, 1990; Porter and Thompson, 1992), these alkaloids may also impart some insect herbivore resistance to symbiotic grasses (Clay and Cheplick, 1989; Siegel et al., 1990; Yates et al., 1989).

In the perennial ryegrass–endophyte symbiosis, indole-isoprenoid lolitrems are potent mycotoxins that cause a disease in livestock commonly referred to as ryegrass staggers (Fletcher and Harvey, 1981). Lolitrem-B and paxilline are

Table 7.1. Secondary Metabolites Identified in Endophyte-Infected Grasses.

***Neotyphodium* Species**	***Balansia* Species**
Ergovaline	Elymoclavine
Ergonovine	Ergonovine
Ergosine	Ergosine
Chanoclavine I	Chanoclavine I
Peramine	Isochanoclavine I
Indole acetic acid	Agroclavine
Ergosterol	Dihydroelymoclavine
Ergosinine	6,7-Secoagroclavine
Cyclopentanoid sesquiterpenoids	Penniclavine
Caffeic acid	3-Indole-acetic acid
p-Coumaric acid	3-Indole ethanol
p-Hydroxybenzoic acid	3-Indole acetamide
Loline	Methyl 3-indole-carboxylate
N-Acetylloline	Ergobalansine
N-Formylloline	Ergobalansinine
N-Acetylnorloline	
Paxilline	
Lolitrem A	
Lolitrem B	
Lolitrem C	
Lolitrem D	

the major alkaloids found in endophyte-infected ryegrass, although minor lolitrems can also be isolated from endophyte-infected tissues (Miles et al., 1994). In addition to being extremely toxic to livestock, the lolitrems may also impart some degree of resistance against the Argentine stem weevil. However, peramine, a specific pyrrolopyrazine alkaloid formed by the synthesis of proline and arginine (Rowan and Gaynor, 1986), is also associated with insect deterrence in endophyte-infected grasses, especially against the Argentine stem weevil. In one of the few surveys of alkaloids from multiple species (Siegel et al., 1990), peramine was found more routinely than any of the other alkaloids except the ergopeptine types, and was identified in plants of *Lolium, Festuca, Poa,* and *Agrostis* spp. This would indicate that peramine synthesis has been favored through evolution because of some unique function.

The pyrrolizidine alkaloids are volatile alkaloids found in high concentrations in endophyte-infected grasses, and were first implicated by Robbins et al. (1972) as a causal factor in poor animal performance. In recent years, it has become apparent that these chemicals are not the primary compounds responsible for animal toxicities, but they have been shown to be involved in insect herbivory (Johnson et al., 1985; Siegel et al., 1990). It is not clear if these alkaloids are produced by the plant in response to the fungus or specifically by the fungus.

BENEFICIAL EFFECTS ON TURFGRASSES — A DEFENSIVE MUTUALISM

Most studies of the effects of endophytes on host fitness have focused on endophyte-infected tall fescue and perennial ryegrass (Siegel et al., 1987; Hill et al., 1991). Much of this work was promoted under the thesis that endophytes are mutualistic with their host plants, a relationship termed a "defensive mutualism" (Clay, 1988). In support of that hypothesis, studies have shown that endophyte-infected grasses are more aggressive (Arechavaleta et al., 1989; Belesky et al., 1989), can acclimate to severe conditions more quickly (West et al., 1993), recover from stressful situations more rapidly (Arechavaleta et al., 1989), and resist insects and fungal pathogens more effectively than non-infected grasses (Rowan and Gaynor, 1986; Siegel et al., 1990; Clarke et al., in press).

Increased insect resistance of endophyte-infected grasses has been well-documented (see review by Breen, 1994). The enhanced insect resistance in these grasses also reflects the mycotoxins produced in the symbiosis (Siegel et al., 1990). The most effective resistance occurs against insects that feed at or near the crown of the plant, the sites of greatest endophyte infection and highest alkaloid content. However, endophytes differ with respect to the alkaloids that they produce (Bacon et al., 1979; Porter et al., 1981), and the degree to which they deter herbivores (Lewis et al., 1993; Siegel et al., 1990). Work by Lewis et al. (1993) demonstrated that locusts were deterred from consuming two species of *Epichloë,* while a third showed no deterrent properties. Owing to this variation, it is reasonable to expect that certain endophytes will prove to be more efficient than others in conferring herbivore resistance to their host. Also, the insect-deterring properties of many endophytes, such as those of sleepy grasses, dronk grasses, and the toxic "huecu grasses" (White, 1988; Bertoni et al., 1993) have not been investigated, and those endophytes might yield a wider range of insect-deterring properties.

In order for turfgrass users to fully realize the biocontrol potential of endophytic fungi, future research must be directed toward identifying, selecting, or manipulating endophytes that produce even higher levels of biocontrol, much like engineering a plant to express a particular trait. One area of endophyte research that can initially lead toward this goal is the screening, selection, and crossing of plant-endophyte populations with elevated levels of specific toxins. Agee and Hill (1994) demonstrated that the production of ergopeptine alkaloids is controlled by host-endophyte interactions, and that low- or high-alkaloid grasses could be developed through traditional recurrent selection techniques. After only two cycles of selection they were able to reduce ergopeptine alkaloid levels in a tall fescue population to less than 100 ppm, where original germplasm contained up to 1000 ppm (N.S. Hill, personal communication). This exciting find suggests that breeders may also be able to select and cross plants with elevated levels of specific toxins such as the insect deterrent peramine. This work is currently under way in our laboratories.

Grasses infected by endophytes also exhibit enhanced drought tolerance relative to noninfected grasses (Read and Camp, 1986; West et al., 1993; Arechaveleta et al., 1989), presumably through alterations in carbohydrate partitioning (Richardson et al., 1992), osmotic adjustment (West et al., 1989), and enhanced root growth (Belesky et al., 1989). It has been shown that simple sugars, including sugar alcohols, and specific amino acids are present in greater concentrations in infected plants (Richardson et al., 1992; Lyons et al., 1990). Unfortunately, not all studies, nor specific clones within a study, have demonstrated a clear improvement in drought tolerance of infected plants compared to noninfected clones (Hill et al., 1996; Richardson et al., 1993; White et al., 1992). Apparently, specific combinations of grasses and endophytes elicit different responses under drought conditions. In one of the most dramatic studies of this phenomenon, Arechaveleta et al. (1989) found that endophyte-infected clones of tall fescue had significantly higher survival rates than endophyte-free clones (75% vs. 0%) following an extended period of desiccation. His study also showed that regrowth following drought stress was significantly improved in infected grasses.

There have been a few isolated descriptions of endophytes enhancing fungal disease resistance of their host (White and Cole, 1985; Funk et al., 1994; Yoshihara et al., 1985; Siegel and Latch, 1991), but this response has been erratic among grass species and specific pathogens. A recent study in New Jersey (Clarke et al., in press) found that endophyte-infected turf plots of Chewings, strong creeping, and hard fescues had significantly better resistance against dollar spot (*Sclerotinia homoeocarpa*) than endophyte-free plots. While the basis of this resistance is presently unknown, the striking results suggest that disease resistance traits may be present in certain endophyte species and, once identified, may be transferred into other useful endophytes. Siegel and Latch (1991) found that *in vitro* growth of several grass pathogens was inhibited by endophytes, suggesting a chemical basis for resistance, and further work in this area is in progress. In Siegel's study, the pathogens most sensitive to endophyte cultures were *Colletotrichum graminicola* and *Rhizoctonia zeae*, with other fungi showing various degrees of sensitivity. Initial work from our laboratories has indicated that specific endophyte isolates from the fine fescues have a high degree of inhibition against *S. homoeocarpa* and *Pyricularia grisea*. Furthermore, chemical extracts from endophyte-infected Chewings fescue seeds strongly inhibit *S. homoeocarpa*, supporting a chemical basis of resistance. Interestingly, the chemicals involved in fungal disease resistance do not appear to be alkaloids (Richardson, unpublished data). These studies suggest that fungal-deterring compounds are present in infected grasses, but further work is necessary to define the significance of these traits.

SOME PROBLEMS LIMITING USE OF ENDOPHYTES IN TURFGRASSES

Potential Toxicities

As we continue to incorporate endophytes into turfgrass varieties, the potential for livestock poisonings must be more rigorously monitored by both the research community and the industry as a whole. The elimination of open-field burning from seed production fields has forced seed growers to manually remove straw and stubble from fields after harvest. At present, most of this straw is baled, removed, and marketed in domestic and foreign markets as a low-quality roughage. Because of the widespread use of endophytes in turf-type tall fescue and perennial ryegrass, reports of animal toxicities associated with seed stubble have occurred (A.M. Craig, personal communication) and ergovaline levels in seed straw are reported as high as 209 ppm (Craig et al., 1994). Therefore, as new varieties are developed with unique strains of endophytes, especially those endophytes with enhanced toxins, the regulation and proper use of seed straw must be considered as a potential source of litigation. A second issue is more relevant to tall fescue than perennial ryegrass or the fine fescues, because tall fescue is truly a multi-use species in the U.S. Increased urban sprawl in the predominant seed growing regions, coupled with increased value of other crops grown in that region, may limit the ability of seed producers to adequately supply seed to both the turf and forage markets. A limited tall fescue supply might dictate the movement of high-endophyte seeds into forage markets, again creating a situation that could result in litigation.

The most promising technologies that can address these issues are the development of vaccines or treatments to ameliorate the toxicosis associated with endophyte-infected grasses. The injection of cattle with a monoclonal antibody that specifically binds ergopeptine alkaloids has been shown to significantly increase serum prolactin levels (Thompson et al., 1993), an indication that metabolic changes associated with fescue toxicosis have been reversed. Other drugs have also shown promise, including ivermectin, a common antiparasite drug used by livestock managers. In a study in Alabama (Bransby et al., 1992), cattle grazing endophyte-infected tall fescue pastures and treated with ivermectin performed equal to animals grazing endophyte-free pastures. While other vaccines and treatments are currently in development, these initial successes would suggest that the potential toxicities of these grasses may soon be overcome through pharmaceutical approaches, and the development of highly deterrent grasses will not be as great a risk.

Another approach that is under investigation is to develop endophyte-infected grasses with little or no toxicological symptoms, while retaining the beneficial effects of the symbiosis. In one recent attempt in New Zealand to produce a nontoxic endophyte-infected cultivar, the identification and incorporation of an endophyte into perennial ryegrass that exhibited limited production of lolitrem B was contradicted by the elevated production of ergopeptine alkaloids. As stated above, breeding efforts to select low alkaloid–producing germplasm is also promising, but the long-term effects of reduced alkaloids on persistence of those grasses remains to be answered.

Absence of Information on the Nature of Host/Endophyte Interactions

Although the importance of grass/fungal endophyte associations is recognized by ecologists (Clay, 1988) and agronomists (Ball et al., 1993) much remains to be learned. In contrast to the information on alkaloids and animal toxicosis, the physiological aspects of the endophyte/grass interactions have not been well characterized in any system. Very little is known regarding the factors important in host colonization or nutrient exchange between plant and fungus. Several studies have indicated endophyte infection can also result in increased plant vigor and confer tolerance to abiotic stresses, unrelated to the reduction in herbivory (Latch and Christensen, 1985; Clay, 1987; Arechaveleta et al., 1989; DeBattista et al., 1990; Rice et al., 1990). The physiological mechanisms which produce these effects are completely unknown. Knowledge of the factors necessary for establishment of effective interactions could be useful in efforts to generate novel grass/fungus combinations with improved agronomic characteristics.

We are investigating the physiological aspects of endophyte/grass interactions with the long-range objective of eventually being able to manipulate agriculturally important interactions. We are using the *Poa ampla* (big bluegrass)/*Neotyphodium typhinum* interaction as a model system (Lindstrom et al., 1993). We have found that in this system, and in other grass/fungal endophyte combinations, a fungal serine proteinase, designated proteinase At1, is produced (Lindstrom et al., 1993; Lindstrom and Belanger, 1994). In some of these interactions this proteinase is amazingly abundant. It was estimated that in the *P. ampla/N. typhinum* interaction proteinase At1, a single fungal protein may represent 1–2% of the total leaf sheath protein (Lindstrom and Belanger, 1994). The enzyme was localized both within membrane vesicles and in the fungal and/or plant cell walls. The abundance and cell wall location suggested that the enzyme may be an important feature in the interaction between the grass and fungus.

We have cloned and sequenced both cDNA and genomic clones for proteinase At1 (GenBank accession no. L76740) (Reddy et al., 1996). Sequence analysis indicated proteinase At1 is a member of the eukaryotic subtilisin-like protease family. The two enzymes most similar to proteinase At1 were from fungal pathogens of nematode eggs (*P. lilacinus*) and insects (*M. anisopliae*) (St. Leger et al., 1992; Bonants et al., 1995). Interestingly, in both of these systems, expression of the homologous protease is hypothesized to be an important feature in pathogenicity. The similarity of proteinase At1 to proteases believed to be virulence factors of pathogenic fungi strengthens the possibility of it being an important factor in the plant/fungus interaction. This is the first protein to be reported which may have a role in the endophytic interaction. To test this possibility directly, we are currently working on targeted disruption of the proteinase At1 gene. A better understanding of endophyte/host interactions may enable us to manipulate the endophyte/grass combination to produce hardier turf cultivars.

Need for Development of Infection Methods

The methods used to incorporate endophytes into turfgrass varieties have varied from selection and backcrossing to direct inoculation of uninfected seedlings or tillers, with each approach having both advantages and limitations (Bacon and White, 1994). Over the past thirty years, breeders of cool-season turfgrasses have inadvertently selected a high percentage of endophyte-infected plants in their programs, primarily because of the positive effects of endophytes on performance. Further screening and selection of endophyte-containing germplasm has led to varieties with greater than 95% endophyte infection levels. However, the absence of an endophyte has not discouraged the selection of breeding materials with desirable traits. Therefore, it is often necessary to introduce endophytes into plants via other methods. Inoculation techniques have been used to introduce endophytes into seedlings (Latch and Christensen, 1985), tiller meristems (O'Sullivan and Latch, 1993), somatic embryos (Kearney et al., 1991), and mature tillers (Belanger, unpublished data), although success levels have generally been very low, ranging from 0 to 50% infection rates. Seedling infection methods have been used most often (Latch and Christensen, 1985), but suffer from the fact that each successful infection will constitute a different plant/endophyte combination. The infection of somatic embryos could theoretically generate clonal plants with different endophyte sources, but this method has been criticized because of the natural somaclonal variation that is inherent with tissue culture techniques (Roylance et al., 1994). The emergence of techniques to inoculate tiller meristems (O'Sullivan and Latch, 1993) and mature tillers (Belanger, unpublished data) provides researchers with the best means to create multiple plant/endophyte combinations such that the expression of desirable traits can be more thoroughly explored. Unfortunately, these techniques are still in their infancy, and their routine use has not been well demonstrated.

Low Endophyte Survival in Seed

As mentioned earlier, the most promising endophytic fungi are those that are exclusively seed-transmitted. Endophyte-infected seed provides an excellent delivery vehicle for moving endophyte-infected plants from the laboratories and breeding nurseries to the production fields and finally the consumer. However, the endophyte will not survive in the seed indefinitely, and storage methods to maintain endophyte viability, such as freezing temperature, are not economically feasible for the seed industry. Storage conditions that accelerate the loss of endophyte in seed include high temperatures and high humidity (Welty et al., 1987). Under improper storage, complete loss of endophyte viability can occur within one year. This loss of viability seems to be more prevalent in the fine fescue species, with endophyte loss occurring as soon as a few months after harvest (C.R. Funk, personal communication). These poor storage capabilities limit the usefulness of endophytes in the fine fescues and create a need to identify endophyte strains that persist better under normal storage conditions.

Recent work in our laboratories has revealed that endophytes from diverse species and geographic locales possess variable fatty acid profiles (Unpublished data). It appears that endophytes that are strictly seed-transmitted have a more complex fatty acid profile than those endophytes that still reproduce sexually through stroma formation. Because fatty acids are utilized as an alternative storage compound in many fungal organisms, we have hypothesized that the additional fatty acids may be important to survival in seed. We are presently screening endophyte isolates from the fine fescue species, with the hopes of identifying specific isolates that possess this complex fatty acid profile. To date, we have identified a single fine fescue isolate that possesses these complex fatty acids, and we are studying other characteristics of this isolate more closely.

Seed Reductions Caused by Stromata Development

The development of stromata on culms of grasses causes reduction in seed production, and may be a source of alarm for users of endophyte-infected turfgrasses. It is through the use of the stroma that endophytes undergo sexual reproduction. Endophytes of tall fescue and perennial ryegrass do not form stromata; however, endophytes in bentgrasses and fine fescues vary in their capacity to produce stromata. Work is needed to identify endophytes that are compatible with these turf species that do not produce stromata and incorporate them into cultivars. We are presently screening germplasm in the Rutgers University fine fescue collections for endophytes that do not form stromata on grasses.

Absence of Endophytes in Bluegrasses and Bentgrasses

The fact that two important turfgrasses — Kentucky bluegrass and creeping bentgrass — lack endophytes is a limitation to the use of endophytes in turf cultivars. However, numerous noncultivated relatives of these turf species do contain endophytes that may be useful in cultivars. An intense effort is being made to locate endophytes that naturally occur in bluegrasses and bentgrasses. These may then be inoculated into Kentucky bluegrass and creeping bentgrass breeding material and incorporated into cultivars. Alternatively, endophyte-infected wild relatives of turfgrasses may be hybridized with those species to transfer both genes for compatibility with the endophyte and the endophyte itself into the turf cultivars. We are attempting both approaches at Rutgers.

CONCLUSIONS

Endophytes hold considerable potential for use in turfgrasses. However, several problems need to be overcome. Among these problems are occurrence of mammalian toxicosis, absence of knowledge on endophyte/grass interactions, absence of efficient procedures for inoculation of endophytes into turf cultivars, limited viability of endophytes in stored seeds, seed reduction due to development of stromata, and absence of endophytes in Kentucky bluegrass and creeping bentgrass.

REFERENCES

Agee, C.S. and N.S. Hill. 1994. Ergovaline variability in *Acremonium*-infected tall fescue due to environment and plant genotype. *Crop Science,* 34:221–226.

Arechaveleta, M., C.W. Bacon, C.S. Hoveland, and D.E. Radcliffe. 1989. Effect of the tall fescue endophyte on plant response to environmental stress. *Agronomy Journal,* 81:83–90.

Bacon, C.W. and J.F. White, Jr. 1994. Stains, media, and procedures for analyzing endophytes. In *Biotechnology of endophytic fungi of grasses.* Bacon C.W. and J.F. White, Jr., Eds. CRC Press, Boca Raton, FL.

Bacon, C.W., J.K. Porter, and J.D. Robbins. 1979. Laboratory production of ergot alkaloids by species of Balansia. *Journal of General Microbiology,* 113:119–126.

Bacon, C.W., J.K. Porter, J.D. Robbins, and E.S. Luttrell. 1977. *Epichloë typhina* from toxic tall fescue grasses. *Applied and Environmental Microbiology,* 34:576–581.

Ball, D.M., J.F. Pedersen, and G.D. Lacefield. 1993. The tall-fescue endophyte. *American Scientist,* 81:370–379.

Belanger, F.C. Rutgers University. Unpublished data.

Belesky, D.P., W.C. Stringer, and N.S. Hill. 1989. Influence of endophyte and water regime upon tall fescue accessions. I. Growth characteristics. *Annals of Botany,* 63:495–503.

Bertoni, M.D., D. Cabral, N. Romero, and J. Dubcovsky. 1993. Endofitos fúngicos en especies sudamericanas de *Festuca* (Poaceae). *Bol. Soc. Argentina Bot.,* 29:25–34.

Bonants, P.J.M., P.F.L. Fitters, H. Thijs, E. den Belder, C. Waalwijk, and J.W.D.M. Henfling. 1995. A basic serine protease from *Paecilomyces lilacinus* with biological activity against *Meloidogyne hapla* eggs. *Microbiology,* 141:775–784.

Bransby, D.I., J.L. Holliman, and J.T. Eason. 1992. Huge weight gain increase from deworming cattle on infected fescue, *Highlights of Agricultural Research — Alabama Agriculture Experiment Station,* 39:9.

Breen, J.P. 1994. *Acremonium* endophyte interactions with enhanced plant resistance to insects. *Annual Review of Entomology,* 39:401–423.

Clarke, B.B., J.F. White, Jr., C.R. Funk, Jr., S. Sun, D.R. Huff, and R.H. Hurley. 1997. Enhanced resistance to dollar spot in endophyte-infected fine fescues. *Plant Disease,* in press.

Clay, K. 1987. Effects of fungal endophytes on the seed and seedling biology of *Lolium perenne* and *Festuca arundinacea. Oecologia* (Berlin), 73:358–362.

Clay, K. 1988. Fungal endophytes of grasses: A defensive mutualism between plants and fungi. *Ecology,* 69:10–16.

Clay, K. and G.P. Cheplick. 1989. Effect of ergot alkaloids from fungal endophyte-infected grasses on fall armyworm (*Spodoptera frugiperda*). *Journal of Chemical Ecology,* 15:169–182.

Clay, K. and A. Leuchtmann. 1989. Infection of woodland grasses by fungal endophytes. *Mycologia,* 81:805–811.

Craig, A.M. Oregon State University, personal communication.

Craig, A.M., D. Bilich, J.T. Hovermale, and R.E. Welty. 1994. Improved extraction and HPLC methods for ergovaline from plant material and rumen fluid. *Journal of Veterinary Diagnostics and Investigations,* 6:348–352.

DeBattista, J.P., J.H. Bouton, C.W. Bacon, and M.R. Siegel. 1990. Rhizome and herbage production of endophyte-removed tall fescue clones and populations. *Agronomy Journal,* 82:651–654.

Fletcher, L.R. and I.C. Harvey. 1981. An association of *Lolium* endophyte with ryegrass staggers. *New Zealand Veterinary Journal,* 29:185–186.

Funk, C.R. Rutgers University. Unpublished data.

Funk, C.R., F.C. Belanger, and J.A. Murphy. 1994. Role of endophytes in grasses used for turf and soil conservation. In *Biotechology of Endophytic Fungi.* C.W. Bacon and J.F. White, Jr., Eds. CRC Press, Boca Raton, FL.

Hill, N.S. University of Georgia, personal communication.

Hill, N.S., D.P. Belesky, and W.C. Stringer. 1991. Competitiveness of tall fescue (*Festuca arundinacea* Schreb.) as influenced by endophyte (*Acremonium coenophialum* Morgan-Jones and Gams). *Crop Science,* 31:185–190.

Hill, N.S., J.G. Pachon, and C.W. Bacon. 1996. *Acremonium coenophialum*-mediated short- and long-term drought acclimation in tall fescue. *Crop Science,* 36:665–672.

Hoveland, C.S. 1992. Importance and economic significance of the *Acremonium* endophytes to performance of animals and grass plant. *Agriculture Ecosystems and Environment,* 44:3–12.

Johnson, M.C., D.L. Dahlman, M.R. Siegel, L.P. Bush, G.C.M. Latch, D.A. Potter, and D.R. Varney. 1985. Insect feeding deterrents in endophyte-infected tall fescue. *Applied and Environmental Microbiology,* 49:568–571.

Kearney, J.F., W.A. Parrott, and N.S. Hill. 1991. Infection of somatic embryos of tall fescue with *Acremonium coenophialum. Crop Science,* 31:979–984.

Latch, G.C.M. and M.J. Christensen. 1985. Artificial infection of grasses with endophytes. *Annals of Applied Biology,* 107:17–24.

Lewis, G.C., J.F. White, Jr., and J. Bonnefont. 1993. Evaluation of grasses infected with fungal endophytes against locusts. *Tests of Agrochemical and Cultivars,* 14:142–143.

Lindstrom, J.T. and F.C. Belanger. 1994. Purification and characterization of an endophytic fungal proteinase that is abundantly expressed in the infected host grass. *Plant Physiology,* 106:7–16.

Lindstrom, J.T., S. Sun, and F.C. Belanger. 1993. A novel fungal protease expressed in endophytic infection of *Poa* species. *Plant Physiology,* 102:645–650.

Lyons, P.C., J.J. Evans, and C.W. Bacon. 1990. Effects of the fungal endophyte *Acremonium coenophialum* on nitrogen accumulation and metabolism in tall fescue. *Plant Physiology,* 92:726–732.

Lyons, P.C., R.D. Plattner, and C.W. Bacon. 1986. Occurrence of peptide and clavine ergot alkaloids in tall fescue grass. *Science,* 232:487–489.

Miles, C.O., S.C. Munday, A.L. Wilkins, R.M. Ede, and N.R. Towers. 1994. Large-scale isolation of lolitrem B and structure elucidation of lolitrem E. *Journal of Agriculture and Food Chemistry,* 42:1488.

O'Sullivan, B.D. and G.C.M. Latch. 1993. Infection of plantlets, derived from ryegrass and tall fescue, with *Acremonium* endophytes, *Second International Symposium on Acremonium/Grass Interactions, New Zealand,* pp 16–17.

Porter, J.K., C.W. Bacon, J.D. Robbins, and D. Betowski. 1981. Ergot alkaloid identification in Clavicipitaceae systemic fungi of pasture grasses. *Journal of Agriculture and Food Chemistry,* 29:653–657.

Porter, J.K. and F.N. Thompson, Jr. 1992. Effects of fescue toxicosis on reproduction in livestock. *Journal of Animal Science,* 70:1594–1603.

Read, J.C. and B.J. Camp. 1986. The effect of the fungal endophyte *Acremonium coenophialum* in tall fescue on animal performance, toxicity, and stand maintenance. *Agronomy Journal,* 78:848–850.

Reddy, P.V., C.K. Lam, and F.C. Belanger. 1996. Mutualistic fungal endophytes express a proteinase that is homologous to proteases suspected to be important in fungal pathogenicity. *Plant Physiology*, 111:1209–1218.

Rehacek, Z. and P. Sajdl. 1990. *Ergot Alkaloids*, Academia, Praha.

Rice, J.S., B.W. Pinkerton, W.C. Stringer, and D.J. Undersander. 1990. Seed production in tall fescue as affected by fungal endophyte. *Crop Science,* 30:1303–1305.

Richardson, M.D. Rutgers University. Unpublished data.

Richardson, M.D., G.W. Chapman, Jr., C.S. Hoveland, and C.W. Bacon. 1992. Sugar alcohols in endophyte-infected tall fescue under drought. *Crop Science,* 32:1060–1061.

Richardson, M.D., C.S. Hoveland, and C.W. Bacon. 1993. Photosynthesis and stomatal conductance of symbiotic and nonsymbiotic tall fescue. *Crop Science,* 33:145–149.

Robbins, J.D., J.G. Sweeny, S.R. Wilkinson, and D. Burdic. 1972. Volatile alkaloids of Kentucky-31 tall fescue seed (*Festuca arundinacea* Schreb). *Journal of Agriculture and Food Chemistry,* 20:1040–1043.

Rowan, D.D. and D.L. Gaynor. 1986. Isolation of feeding deterrents against Argentine stem weevil from ryegrass infected with the fungal endophyte *Acremonium loliae. Journal of Chemical Ecology,* 12:647–658.

Roylance, J.T., N.S. Hill, and W.A. Parrot. 1994. Detection of somaclonal variation in tissue culture regenerants of tall fescue. *Crop Science,* 34:1369–1372.

St. Leger, R.J., D.C. Frank, D.W. Roberts, and R.C. Staples. 1992. Molecular cloning and regulatory analysis of the cuticle-degrading-protease structural gene from the entomopathogenic fungus *Metarhizium anisopliae. Eur. J. Biochem.*, 204:991–1001.

Siegel, M.R. and G.C.M. Latch. 1991. Expression of antifungal activity in agar culture by isolates of grass endophytes. *Mycologia,* 83:529–537.

Siegel, M.R., G.C.M. Latch, L.P. Bush, F.F. Fannin, D.D. Rowan, B.A. Tapper, C.W. Bacon, and M.C. Johnson. 1990. Fungal endophyte-infected grasses: Alkaloid accumulation and aphid response. *Journal of Chemical Ecology,* 16:3301–3315.

Siegel, M.R., G.C.M. Latch, and M.C. Johnson. 1987. Fungal endophytes of grasses. *Annual Review of Phytopathology,* 25:293–315.

Thompson, F.N., N.S. Hill, D.L. Dawe, and J.A. Stuedemann. 1993. The effects of passive immunization against lysergic acid derivatives on serum prolactin in steers grazing endophyte-infected tall fescue, *Second International Symposium on Acremonium/Grass Interactions, New Zealand,* pp. 135–137.

Welty, R.E., M.D. Azevedo, and T.M. Cooper. 1987. Influence of moisture content, temperature, and length of storage on seed germination and survival of endophytic fungi in seeds of tall fescue and perennial ryegrass. *Phytopathology,* 77:893–900.

West, C.P., E. Izekor, K.E. Turner, and A.A. Elmi. 1993. Endophyte effects on growth and persistence of tall fescue along a water-supply gradient. *Agronomy Journal,* 85:264–270.

West, C.P., D.M. Oosterhuis, and S.D. Wullschleger. 1989. Osmotic adjustment in tissues of tall fescue in response to water deficit. *Environmental and Experimental Botany*, 30:149–156.

White, J.F., Jr. 1988. Endophyte-host association in forage grasses. XI. A proposal concerning origin and evolution. *Mycologia,* 80:442–446.

White, J.F., Jr. and K. Camp. 1995. A study of water relations of *Epichloë amarillans* White, an endophyte of the grass *Agrostis hiemalis* (Walt.) B.S.P. *Symbiosis,* 18:15–25.

White, J.F., Jr. and G. T. Cole. 1985. Endophyte-host associations in forage grasses. III *In vitro* inhibition of fungi by *Acremonium coenophialum. Mycologia,* 78:102–107.

White, R.H., M.C. Engelke, S.J. Morton, J. Johnson-Cicalese, and B.A. Ruemmele. 1992. *Acremonium* endophyte effects on tall fescue drought tolerance. *Crop Science,* 32:1392–1396.

Yates, S.G., J.C. Fenster, and R.J. Bartelt. 1989. Assay of tall fescue seed extracts, fractions, and alkaloids using the large milkweed bug. *Journal of Agriculture and Food Chemistry,* 37:354–357.

Yates, S.G., R.J. Petroski, and R.G. Powell. 1990. Analysis of loline alkaloids in endophyte-infected tall fescue by capillary gas chromatography. *Journal of Agriculture and Food Chemistry,* 38:182–185.

Yates, S.G. and R.G. Powell. 1988. Analysis of ergopeptine alkaloids in endophyte-infected tall fescue. *Journal of Agriculture and Food Chemistry,* 36:337–340.

Yoshihara, T., S. Togiya, H. Koshino, S. Sakamura, T. Shimanuki, T. Sato, and A. Tajimi. 1985. Three fungitoxic cyclopentanoid sesquiterpenes from stromata of *Epichloë typhina. Tetrahedron Letters,* 26:5551–5554.

Part 3

Genes with Potential for Turfgrass Improvement

Chapter 8

Molecular Identification of Cold Acclimation Genes in, and Phylogenetic Relationships Among, *Cynodon* Species

M.P. Anderson, C. Taliaferro, M. Gatschet, B. de los Reyes, and S. Assefa

INTRODUCTION

Bermudagrass, *Cynodon dactylon* L. Pers., is a warm-season perennial sod-forming species widely used for turf, forage, and soil stabilization. Its natural distribution and use is primarily in warmer climatic regions, though it is distributed throughout the world between the latitudes of 45°N and 45°S (Harlan et al., 1970). It is found on every continent and in the islands of the Pacific Ocean. The benefits of turf bermudagrass include a high level of wearability, salt and drought tolerance, and many pleasing aesthetic qualities (Gatschet, 1993). In the United States, turf bermudagrass is used primarily in the southeast, though its use extends throughout the southern states into the more temperate central region of the country. Its southern localization is mainly due to the limited degree of tolerance to the low minimum winter temperatures found in more northern climates. Within the zone of adaptation severe damage to bermudagrass can result from occasional freezing conditions (Fry, 1990). Published killing temperature thresholds for acclimated bermudagrass vary from –4 to –17°C, indicating a significant degree of genetic variation for this trait (Fry, 1990; Anderson et al., 1988; Anderson et al., 1993). Increased cold hardiness in turf bermudagrass cultivars is desirable to minimize or prevent freeze injury, which periodically results during severe winters. Such injury disrupts use of the turfed area and is costly to replace. Replacement value of this turf species in Oklahoma alone has been estimated at approximately 1.3 billion dollars (Gatschet, 1993).

The ability of plants to acclimate their metabolism to freezing stress is a complex polygenic phenomenon (Thomashow, 1990; Sakai and Larcher, 1987). Acclimation enables plant cells to adapt to a series of cellular and biochemical stresses that accompany the freezing process. Acclimation is gradual, in that it begins as fall temperatures decrease and continues until maximum cold hardiness is achieved. In order to improve cold tolerance in bermudagrass we must first understand the freezing and acclimation process(es) at the cellular and biochemical levels.

Freezing usually first occurs outside the plasma membrane, resulting in a lowering of the chemical water potential. The lowered water potential causes water to flow out of the cell due to the osmotic effects mediated by the plasma membrane. Intracellular dehydration, increases in salt concentrations, and an increase in toxic metabolites result from the reduction of water activity within the cell (Steponkus, 1984). The ability of the plasma membrane to withstand stresses due to extracellular freezing is central to the cold acclimation process. Many investigations have focused on the plasma membrane as the critical site for cold acclimation studies. Rupture of the plasma membrane is a primary event that precedes cell death. The plasma membrane appears to function in two ways: (1) as a barrier against the propagation of extracellular freezing into the cell, and (2) as an osmometer that allows for the outflow of intracellular water to reestablish thermodynamic equilibrium (Steponkus, 1984). Because of its central role, most investigations have focused on the lipid composition of the plasma membrane as a critical parameter for cold acclimation (Uemura et al., 1995). Other studies have focused on changes in the structure and morphology of the plasma membrane, and its effect on membrane stability (Webb et al., 1994; Webb and Steponkus, 1993). Effects not directly associated with the plasma membrane are also important to the acclimation process. Dehydration or increase in intracellular salt concentration may disrupt normal cellular metabolism. Late-embryogenesis abundant proteins and dehydrins are thought to be associated with freeze-induced dehydration stress (Hincha et al., 1990). How these proteins function as protectants is unclear. Electron leakage from membrane-bound electron transport chains results in oxidative stress under low-temperature conditions. Acclimation may require the induction of oxidative enzymes to counteract these effects (McKersie et al., 1994; Hausladen and Alscher, 1994). Low temperatures may destabilize critical enzymes. Induction of cryoprotection proteins may protect enzymes against this cold-temperature inactivation (Lin and Thomashow, 1992; Sieg et al., 1996). Reduction in freezing may be in response to proteins that inhibit ice nucleation (Griffith et al., 1992), and increases in carbohydrate concentrations may serve as cryoprotectants against freeze-induced damage (Leborgne et al., 1995; Antikainen and Pihakaski, 1994). The above are just a sampling of the many effects that result from the cold acclimation and freezing processes. At this point we are just beginning to glimpse the biochemical and molecular outlines of cold acclimation mechanisms in plants.

Most studies involving cold acclimation concentrate on two distinct approaches: (1) isolation of proteins or genes that are differentially expressed using two-dimensional electrophoresis, differential screening of cDNA clones, or subtractive hybridization, and (2) biochemical studies that seek to identify the primary and secondary events associated with cold-induced damage and cold acclimation. We have chosen to first isolate cold acclimation genes from bermudagrass crown tissues using a new highly efficient molecular technique. Once a number of genes have been isolated and identified, we plan to pursue biochemical characterization of the gene products in order to learn more about their function. As more genes are identified and characterized, and as the biochemical responses to cold temperatures are better defined, the pieces of the larger puzzle will begin to take shape, and connections between the molecular and biochemical approaches will emerge into a more coherent picture.

For the last four years we have focused our efforts on isolating differentially expressed cold acclimation proteins using two-dimensional electrophoresis. One cold-regulated (*cor*) protein has been identified as a chitinase (Gatschet, 1993; Gatschet et al., 1994; and Gatschet, 1996). This approach has been technically demanding, and success was not guaranteed from the beginning, but since then we have shifted our focus toward the use of a more powerful molecular approach known as the differential display to more efficiently isolate *cor* genes.

The differential display procedure was first introduced in 1992 by Liang and Pardee, 1992 with numerous modifications thereafter (Bauer et al., 1993; Liang et al., 1995). Essentially, the technique is PCR-based, utilizing two distinct primers, termed the anchored and arbitrary primers, to fractionate the total expressed mRNA population into distinct subsets. By combining different arbitrary and anchored primers we can separate and theoretically isolate all mRNA using a series of high-resolution polyacrylamide gels. By comparing changes in banding patterns across treatments, differentially expressed genes can be identified. Isolation and identification of gene product requires gel elution, PCR amplification, and subcloning into a plasmid vector, followed by nucleic acid sequencing. Having experienced earlier techniques, such as the two-dimensional electrophoresis, we have found the differential display to be the easiest and quickest approach to isolate differentially expressed genes. Here we report on our success with two-dimensional electrophoresis in isolating a *cor* protein, and compare it with our current progress using the differential display.

In addition to our work with cold acclimation, we are also interested in characterizing the genetic diversity among *Cynodon* taxa and ecotypes within taxa. Harlan and de Wet erected a taxonomic structure for *Cynodon* using morphological markers, cytogenetic traits, and crossing ability (Harlan et al., 1970a,b; de Wet and Harlan, 1970). From this analysis they separated the genus into nine different species (Harlan et al., 1970). *Cynodon dactylon* is the most cosmopolitan species of all, having a tremendous diversity in growth range and habit. Most commercial cultivars of bermudagrasses are from C. *dactylon*, *C. transvaalensis* or hybrids of the two species. All seed-propagated turf bermudagrass cultivars are of the taxon *C. dactylon var. dactylon.* The widespread use of only a few triploid (2n=3x=27) vegetatively propagated cultivars — and a few genetically similar seed-propagated cultivars — renders the species genetically vulnerable (Taliaferro, 1995). Furthermore, the current bermudagrass collection in the National Plant Germplasm system is not sufficiently broad or well-characterized to allow for better exploitation of genetic diversity. Of the 484 accessions, 411 are *C. dactylon*. Some *Cynodon* species are not represented at all (Taliaferro, 1995). Risk reduction to diseases and insects through enhancement of genetic diversity will require techniques that are capable of accurately assessing genetic diversity in current bermudagrass collections. Enhancement of bermudagrass characteristics through conventional breeding will be greatly benefited by a knowledge of the genetic interrelationships within the genus.

Recent developments have yielded a technique known as DNA amplification fingerprinting (DAF) that is capable of detecting many DNA polymorphisms (Caetano-Anollés et al., 1991). This technique is PCR-based, utilizes short 5 to 8 base pair oligonucleotide primers, is fast and simple to perform, uses no radionucleotides, and is sensitive to low copy number DNA. The data derived from DAF protocols can be used to construct phylogenetic trees using computer-based parsimony analysis. In this chapter we are reporting on our progress using the DAF procedure to ascertain the genetic variability within the most cosmopolitan bermudagrass species, *C. dactylon*. Later we plan to expand the use of the DAF technique to cover a collection of 60 bermudagrasses inclusive of all nine species in our OSU collection.

METHODS

Plant Culture

Midiron and Tifgreen turf bermudagrasses were grown under controlled environmental conditions of 2/8°C day/night for treated and 28/24°C day/night for control tissues. Midiron possesses a high level and Tifgreen an intermediate level of cold tolerance. Plants were grown in Cone-tainers or six inch pots supplemented with macro- and micronutrients. Crown tissues were harvested at 4°C with care to minimize degradation of proteins (Gatschet, 1993; Gatschet, 1996) or RNA.

Two-Dimensional Electrophoresis of Crown Tissue Proteins

Crown tissue prior to extraction was radiolabeled for 16 hr with ^{35}S-methionine and ^{35}S-cysteine Trans ^{35}S label reagent (ICN Biochemicals, Irvine, California). Extraction of radiolabeled proteins was performed according to Damerval et al., 1986; Gatschet, 1993; and Gatschet et al., 1996. Treatments and extraction were replicated at least two times. Protein concentration was determined using the DC Bio Rad protein assay (Bio Rad, Hercules, California). Quantitation of radioactivity was performed by liquid scintillation spectroscopy. Proteins were dissolved in Isoelectric focusing (IEF) solubilization buffer and diluted so that each sample was equal with respect to protein concentration and specific activity (Gatschet et al., 1994b). IEF was carried out using 4% T, 5.5% C with a mixture of 3-10, 5-7, and 7-9 ampholytes in a 1:2:2 volume ratio from Serva, Serva and Bio Rad, respectively. The catholyte was 0.1 *M* NaOH and the anolyte 6 m*M* phosphoric acid. Total protein loaded was 32 μg in 20 μL of UKS buffer. Samples were subjected to electrophoresis using 500 V for 10 min and 750 V for 3.5 hr using the Bio Rad Mini Protean II system. Gels were extruded (Porter et al., 1992) and equilibrated for 45 min, and stored in parafilm strips at –70°C. The second dimension gel consisted of a 12% T, 2.5% C separator according to Laemmli (1970) and Bio Rad instruction manual. Marker proteins were used in some gels to calibrate with respect to pI and molecular weight. Second dimension was run for 45 min at 200 V. Gels were stained with silver (Blum et al., 1987), photographed, destained (Kulsar and Prestwich, 1988), impregnated with fluor (Porter and Gatschet, 1992), dried (Porter and Gatschet, 1992), and autoradiographed using Kodak X-Omat AR film at –70°C. Exposure times were for 3 to 4 weeks. Silver-stained gels were autoradiographed and then scanned using a PDI densitometer (Huntington Station, New York).

27 kD Protein Isolation and Sequencing

Isolation of 27 kD protein was performed by preparative two-dimensional electrophoresis according to Gatschet (1996). Gels were stained with Coomassie Brilliant Blue R250 (Rosenfeld et al., 1992), and the 27 kD protein was excised. The gels were destained and silver-stained to confirm 27 kD protein isolation. A total of 35 spots were excised and frozen at –70°C from an equal number of gels for microsequencing. Thawed gels were subjected to trypsin digestion (Rosenfeld et al., 1992). Amino acid sequencing was performed in the University of Oklahoma Health Sciences Sequencing Unit according to Gatschet et al. (1996). Sequence alignment and identification were performed using the BLAST search program of the National Center for Biotechnology Information database.

Isolation of *Cor* genes

Plants were treated and crowns harvested as indicated above, except a 2-day, 28-day cold acclimation followed by a 2-day deacclimation treatment (28/24°C, day/night) was used. Total RNA was isolated from freshly isolated crown tissue using a modified Logemann procedure (Logemann et al., 1987). The usual precautions in working with RNA were used to ensure good quality starting material (Farrell, 1993). Tissue was finely ground in liquid nitrogen and then RNA was extracted using a phenol:chloroform:isoamyl alcohol (25:24:1) solution three times. RNA was precipitated overnight using ethanol-acetic acid precipitation at –21°C followed by three washes in 3 *M* sodium acetate and 70% ethanol. The pellet was resuspended in DEPC-treated water and precipitated for 4 hr at 4°C with 4 *M* LiCl. The RNA was centrifuged and the pellet washed twice with 70% ethanol, dried under vacuum for 10 min, and dissolved in DEPC water. RNA was quantified using 260 nm and purity assessed by agarose electrophoresis (Sambrook et al., 1989) and 260/280 ratios (Sambrook et al., 1989). RNA was treated with DNAse according to instructions from the Message Clean Kit from GenHunter Corp. The differential display was performed using a MJ Research PTC-200 thermalcycler according to the GenHunter Corp. RNAmap kit instructions. Radiolabeling was performed using either ^{35}S or ^{33}P isotopes. Separation of gene fragments was performed according to the instructions from the Bio Rad Sequigene II instructions. Autoradiography was performed using Amersham Hyperfilm MP X-ray film for 2 to 4 days at –80°C. Gene fragments of interest were excised and eluted using the QIAEX II DNA extraction kit (QIAGEN Inc., Chatsworth, CA). Partial cDNAs were PCR amplified according to RNA map kit instruction and cDNA fragment subcloned into a Invitrogen TA cloning vector (Invitrogen, Carlsbad, CA). Subcloned plasmids will be submitted after alkaline lysis isolation (Sambrook et al., 1989) to the Sequencing Facility on campus for nucleic acid sequencing.

Phylogenetic Analysis of *Cynodon dactylon* Lines

Cynodon dactylon lines were grown in the greenhouse facility from seed or clonally propagated stocks in our collection at Oklahoma State University. Genomic DNA was extracted from 4 g of leaf tissue using phenol-chloroform extraction protocol modified from Sastry et al. (1995). Tissue was ground in liquid nitrogen to a fine powder, transferred into extraction buffer, and incubated at 65°C for 20 min. The DNA was purified once using phenol:chloroform (1:1) and then by chloroform:isoamyl alcohol (24:1). The DNA was precipitated using isopropanol at –20°C, mixed gently and swirled to precipitate and aggregate the DNA, incubated at –20°C for 15 min and recovered by spooling onto a glass rod. The DNA was dried, suspended in TE pH 8.0 and treated with RNAse at 37°C for 2 hr. The DNA was phenol:chloroform:isoamyl alcohol

(25:24:1) extracted and precipitated with absolute ethanol. The pellet was washed with 70% ethanol, dried in vacuum and resuspended in TE pH 8.0. Concentration was determined by fluorescence. Electrophoretic separation of DNA fragments was performed using a Bio Rad mini Protean II with modifications (Caetano-Anollés et al., 1991). Rather than use the thin 0.4 mm gel format, we used the conventional 1 mm gels without polyester backing. In addition, we reduced the TEMED and ammonium persulfate concentrations by half. After electrophoresis, the gels were silver-stained according to Bassam and Caetano-Anollés (1993), dried between two sheets of cellulose acetate and set in a drying frame. Analysis of banding pattern was performed, scoring for the presence or absence of visually apparent bands. The phylogenetic analysis was performed using PAUP 3.0 (David Swofford, 1993). The data were entered as unweighted, unrooted, and unordered characters. An exhaustive search using the branch and bound algorithm to identify the minimal tree was performed. Its significance was addressed with 100 bootstrap replications using a 50% consensus tree calculation.

RESULTS AND DISCUSSION

Cor Proteins

Two-dimensional electrophoretic analysis has been used successfully to isolate differentially expressed proteins in response to a variety of treatments and conditions. We used this technique to isolate and identify proteins that are differentially expressed in cold-acclimated crown tissues of bermudagrass (Gatschet, 1996) (Figure 8.1). Differential expression of *cor* proteins was clearly evident in gels visualized by silver staining and fluorography. Expression of a 34 kD neutral protein designated M1 and M2 in Midiron and T1 and T2 in Tifgreen was diminished by the cold acclimation process. In addition, cold acclimation increased the expression of a number of basic proteins in the molecular weight range of 20 to 28 kD (low-molecular-weight basic, LB), 35 to 37 kD (intermediate-molecular-weight basic, IB) and 46 to 57 kD (high-molecular-weight basic, HB). Moreover, several acidic proteins were more highly expressed in cold-acclimated crown tissues with molecular weights ranging from 32 to 37 kD (intermediate-molecular-weight acidic, IA) and 14 to 15 kD (low-molecular-weight acidic, LA). Proteins with molecular weights approximating 34 kD were more abundant in silver-stained gels in both Midiron and Tifgreen crowns after the 26-day acclimation period compared to nonacclimated crowns, but for unknown reasons were not being currently synthesized in detectable amounts at the end of the acclimation period. Low-molecular-weight basic proteins in acclimating crown tissues were actively synthesized to a much greater degree in Midiron than in Tifgreen. However, these spots were not readily apparent on the silver-stained gels, indicating that they were not very abundant.

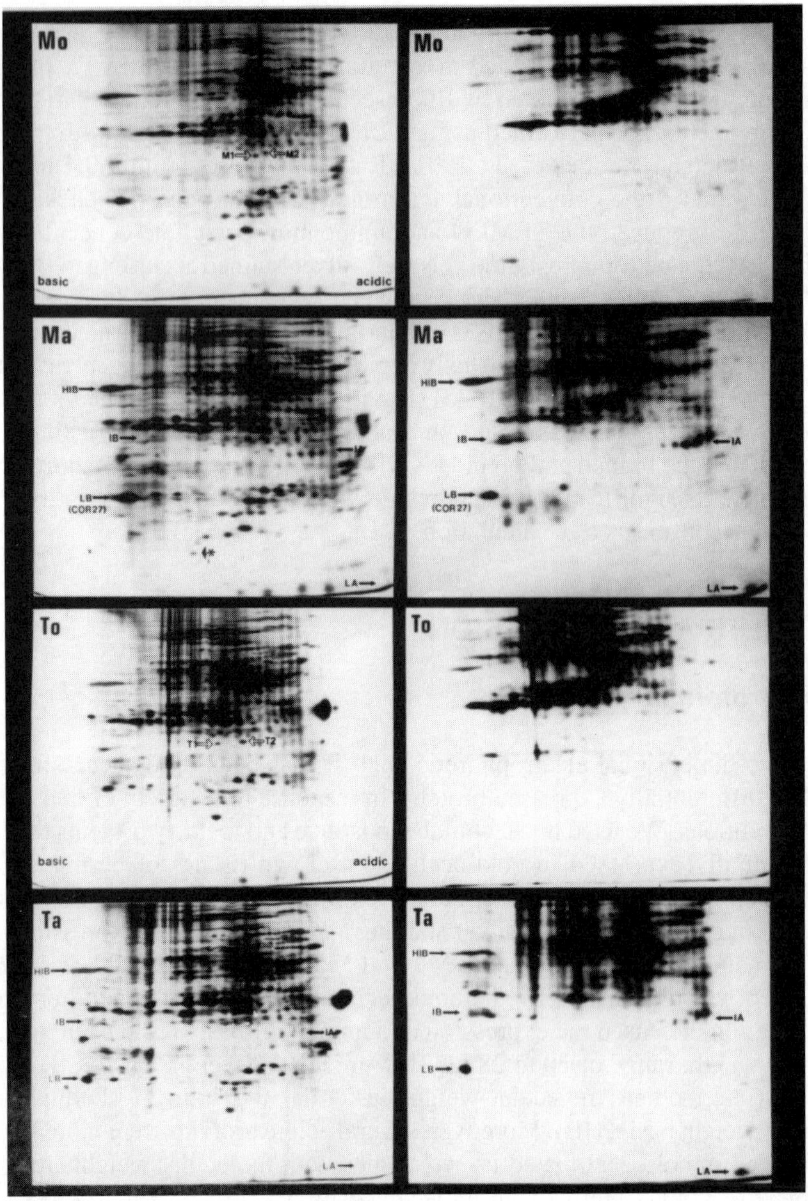

Figure 8.1. Two-dimensional electrophoresis gels from acclimated (Ma,Ta) and control (Mo,To) crown tissues of Midiron (M) and Tifgreen (T). Gels on the left are silver-stained, while those on the right are autoradiograms of gels subjected to fluorography. HIB=High-molecular-weight basic proteins, IB=Intermediate-molecular-weight basic proteins, LB=Low-molecular-weight basic proteins, IA=Intermediate-molecular-weight acidic proteins, and LA=Low-molecular-weight acidic proteins. M1, M2, T1, T2 = 34 kD proteins showing diminished expression in response to treatment.

Others have identified differentially expressed proteins in response to cold temperatures using one- or two-dimensional electrophoresis (Gatschet et al., 1994b; Guy and Haskell, 1987; Guy and Haskell, 1988; Perras and Sarhan, 1989). Comparisons of molecular weights as well as pIs with specific *cor* proteins expressed in other species may be useful in identifying proteins that are conserved. A 33 kD (pI 5.2) wheat seedling protein (Perras and Sarhan, 1989), diminished in response to cold treatment, may be homologous to the bermudagrass 34 kD protein referred to above. The 18 kD bermudagrass *cor* protein was similar in size and charge to a 17.6 kD alfalfa protein designated CAS18 (Wolfraim et al., 1993). Several spinach proteins that respond to cold treatment and have an acidic pI were electrophoretically similar to the intermediate acidic bermudagrass proteins (Guy and Haskell, 1988). Further research is necessary to better establish the conservation of specific *cor* proteins using amino acid sequencing and immunological-based comparisons.

We decided to focus on the 27 kD basic bermudagrass protein for further isolation and characterization because it was more actively synthesized and found in greater amounts in both Midiron and Tifgreen crown tissues. This is important since the probability of successfully isolating a protein from two-dimensional polyacrylamide gels is directly related to the amount of extractable protein. Other differentially expressed proteins, such as the HIB, IB, or several of the LB proteins, may also play a significant role in the cold acclimation process, but would be much more difficult to isolate. Accordingly, the 27 kD protein was eluted, enzymatically cleaved into six fragments ranging from 5 to 12 amino acid residues, and sequenced (Table 8.1). Only one of the residues from the sixth fragment was ambiguous with respect to amino acid composition. Searches using the BLAST program of the NCBI protein sequence databases and all amino acid sequences from the six fragments clearly indicated that the 27 kD protein is a chitinase (E.C. 3.2.1.14) (Table 8.2). In fact, the 63 highest scoring matches showed significant homology to plant chitinases or chitinase precursors (data not shown). A 229 amino acid residue from tobacco class II chitinase had the highest sequence homology of them all (Payne et al., 1990). The highest sequence matches were from class II, while a few were from class I and IV chitinases. Further division of plant chitinases into b class (barley chitinase) corresponding to class I, II, and IV, and h type (hevamine type) corresponding to class III and V has been proposed (Beintema, 1994). Class b and h chitinases show very little interclass homology to each other (Beintema, 1994). These results clearly support the contention that the bermudagrass 27 kD COR protein is a type b, class II chitinase.

This leads to the intriguing question concerning the potential involvement of the 27 kD bermudagrass chitinase with respect to the cold acclimation. Chitinases are known primarily for their protective role against fungal or bacterial pathogens (Graham and Sticklen, 1994) and defense reactions. Transgenic plants with overexpressing constructs of other plant chitinases or glucanases show a higher degree of resistance compared to nontransformed plants (Lin et al., 1995; Neuhaus et al., 1992). Chitinases act to degrade cell walls of ingressing

Table 8.1. Amino Acid Sequence of Six Fragments Obtained by Tryptic Digestion of the 27 kD Bermudagrass Protein.

Sequence Number	Amino Acid Sequence
1	FYTYEAFTR
2	NLVGNPNLVATD
3	IINGGVE
4	GPIQLTHR
5	AIWFN
6	INLD?YN

? represents ambiguous residue.

Table 8.2. Amino Acid Sequence Homology to Known Plant Chitinases.

Species	Class	% Residue Identity	Reference
Tobacco PR-Q	II	79	(Payne, Ahl et al., 1990)
Tomato Chi3	II	74	(Danhash, Wagemakers et al., 1993)
Rye seed RSC-c	II	72	(Yamagami and Funatsu, 1993)
Tobacco CHN50	I	66	(Shinshi, Mohnen et al., 1987)
Rapeseed ChB4	IV	49	(Rasmussen, Giese et al., 1992)

plant pathogens (Benhamou et al., 1990; Chang et al., 1995; Arlorio et al., 1992). Products released from the cleavage of cell wall material by fungal pathogens are known to act as elicitors for induction of plant chitinases (Fukuda and Shinshi, 1994). Chitinase proteins are extruded into the extracellular matrix, where they contact invading pathogens. Vacuolar chitinases are thought to slow invading pathogens upon release from the cell by pathogen-induced lysis. Chitinases are typically small proteins with molecular weights ranging from 25 to 35 kD, and are expressed as multiple isozymes with basic and acidic pIs (Graham and Sticklen, 1994). They are found at low constitutive levels in most plant tissues, and are inducible by a wide range of factors, such as microbial infections (Kim et al., 1994), wounding (Chang et al., 1995), ethylene (Roby et al., 1991), oligosaccharide fragments from cell walls (Fukuda and Shinshi, 1994), cold temperatures (Tronsmo et al., 1993), and toxic chemicals (Hauschild, 1993). The large array of inducible responses suggests that chitinases may function in general plant stress mechanisms. Few, if any, endogenous functions have been demonstrated for plant chitinases (Graham and Sticklen, 1994).

Chitinases are induced by low temperatures in other plant species. Rye chitinases were induced in response to pathogen infection and cold temperatures, and these proteins exhibited antifreeze activity (Hon et al., 1995). The antifreeze activity is thought to be effective in decreasing ice nucleation in the extracellular matrix (Hon et al., 1995; Griffith et al., 1992), as well as protecting sensitive enzymes and metabolic components against cold temperature inactivation (Kazuoka and Oeda, 1994; Lin and Thomashow, 1992). This suggests an apparent dual role for chitinases in protecting against both pathogens and low temperature stress (Hon et al., 1995). This dual role may be an evolutionary response to protect plants from pathogen infection when temperatures decrease. The most serious fungal disease in bermudagrass is spring deadspot, caused principally in North America by *Leptosphaeria korrae* J.C. Walker & A.M. Sm, *Ophiosphaerella herpotricha* J.C. Walker, and *Gaeumannomyces graminis* (Sacc.) Arx & D. Olivier *var graminis* (McCarty et al., 1990; Smiley et al., 1992). This soilborne disease is active primarily during cool fall periods in both roots and crown tissues. The relationship between spring deadspot and the 27 kD protein warrants further study.

Cor Genes

Many major technical advances in molecular biology have occurred since we began to isolate proteins using two-dimensional electrophoresis. Recently, a procedure known as the differential display was developed that greatly facilitates this process. Since 1990 there have been over 200 references concerning this technique, most related to mammalian species. The differential display has been used extensively in cancer research to identify differentially expressed genes in breast cancer (Liang and Pardee, 1992; Watson and Fleming, 1994), ovarian cancer (Mok et al., 1994), a potential tumor suppresser gene (Sager et al., 1993), and many more medically related genes. The display is just beginning to be used in plant science, where investigators have identified genes related to freezing tolerance in wheat (Sutton et al., 1995), cotton fiber–specific cDNAs (Song et al., 1995), tomato fruit development (Tieman and Handa, 1996), sucrose regulated genes in rice (Tseng et al., 1995), cDNA induced by gibberellic acid in barley aleurone layers (Rochester et al., 1995), dormancy-related genes in oats (Johnson et al., 1995), and ozone-induced cDNAs in arabidopsis (Sharma and Davis, 1995).

We have chosen to use the differential display to isolate *cor* genes rather than attempt to isolate the other *cor* proteins that are in low abundance in bermudagrass crown tissue. The display procedure has many advantages over previous techniques (Liang and Pardee, 1992). It is much more rapid than two-dimensional electrophoresis, differential screening of cDNA libraries, or subtractive hybridization procedures. Partial cDNAs can be isolated within a week, compared to months for the above mentioned techniques. It took us close to a year to isolate enough chitinase for amino acid sequencing, while in comparison it took only 8 months to isolate 31 differentially expressed bands. The technique is not prone to technical difficulties, as often occurs with two-dimensional electrophoresis, especially with low abundance proteins. The display utilizes some of the most tried and true molecular techniques, making it accessible to most investigators with minimum training in molecular biology. The procedure has been optimized in commercially available kits, further adding to the convenience of the display. The display shows a much higher resolving power than other techniques. The maximum number of resolvable spots on a two-dimensional gel ranges from 300 to 2000, and the display was shown to resolve up to 38,000 bands representing the presumed total mRNA population (Bauer et al., 1993). The display is able to detect low copy and even single copy number genes with great sensitivity. cDNA fragments are more easily eluted and sequenced from the display than are proteins from two-dimensional electrophoresis gels.

To date, we have isolated 31 bands from acclimated, deacclimated, and control crown tissues using the differential display. Furthermore, we have PCR amplified and subcloned 24 of them into the Invitrogen TA cloning vector. These bands normally range between 200 and 400 base pairs in length. Seventeen of these have been PCR amplified and subcloned in preparation for sequencing. Of the 31 bands, six were down-regulated and 25 up-regulated. Knowledge of down-regulated gene expression may be just as significant to our objectives of understanding the cold acclimation mechanism as that of up-regulated gene expression. An example of a typical display gel is presented in Figure 8.2. One down-regulated gene is clearly induced after 2 and 28 days of treatment compared to control. After deacclimation, the down-regulated band returned to control levels. The display also revealed a major up-regulated cDNA. The up-regulated cDNA increased in response to cold temperature after the 2- and 28-day cold treatments and then returned to control levels after the 2-day deacclimation period. The band intensity of both cDNAs closely follows a pattern of expression consistent with both up- and down-regulation, respectively. We are currently sequencing the partial cDNA fragments to determine sequence homology with other genes in the database. We believe that subtle outlines of the cold acclimation mechanism will begin to emerge as the numbers of differentially expressed cDNA are identified. We plan to select the most intriguing cDNAs for further characterization at the biochemical and cellular level. *Cor* genes isolated by the differential display will be tested for increased expression using a wide range of bermudagrass germplasm with varying ability to cold acclimate. Those *cor* genes that best correlate with cold acclimation ability may be used as selectable markers to enhance breeding efficiency for cold acclimation in our bermudagrass germplasm.

Phylogenetic Relationships

Phylogenetic analysis of bermudagrass cultivars within *Cynodon dactylon* was performed using the DAF procedure (Caetano-Anollés et al., 1991). Ten

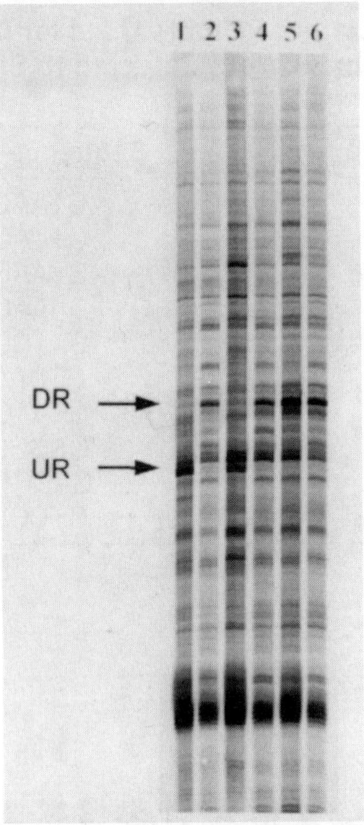

Figure 8.2. Differential display gel showing banding pattern for bermudagrass crown tissues exposed to two-day acclimation treatment (Lane 1), control for two-day acclimation treatment (Lane 2), 28-day acclimation treatment (Lane 3), control for 28-day treatment (Lane 4), two-day deacclimation treatment after 28-day acclimation (Lane 5) and control for two-day deacclimation treatment (Lane 6). DR identifies a cDNA band that is apparently down-regulated, while UR identifies a cDNA band that is up-regulated.

arbitrary primers (Table 8.3) were used to score 292 bands, or 29.2 bands per primer. A sample silver-stained gel is shown in Figure 8.3. Parsimony and bootstrap analysis was performed to produce the consensus phylogenetic tree using the PAUP 3.0 program. Results are presented in Figure 8.4, along with estimates of branch lengths. Correlations between the phylogenetic analysis and knowledge of genetic and geographic relationships appears to confirm the validity of the technique within the *Cynodon dactylon*. Accessions A9958 and A9959 from Italy and Serbia, respectively, represent the common temperate race of *C. dactylon var. dactylon*, and are shown to be closely linked according to the results generated by the DAF technique. Beijing and Tifton 10 are of similar geographic origin, and according to our results are genetically linked. The close relationship between Texturf 10 and Tifton 10 was also documented

Table 8.3. Arbitrary Octamer Primers Used for DAF Analysis of Nine *C. Dactylon* Lines. All Primers Were Selected to Show a GC Content of at Least 50%.

Primer Code	Sequence (5′ to 3′)
1924	GTAACCCC
1985	GTTACGCC
1986	GTAACGCC
1949	GAAACGCC
2273	GATCGCAG
1952	GTCAATTC
2286	GTATCGCC
2287	GAGCCTGT
2288	GAGGGTGG
2289	GTCCAATC

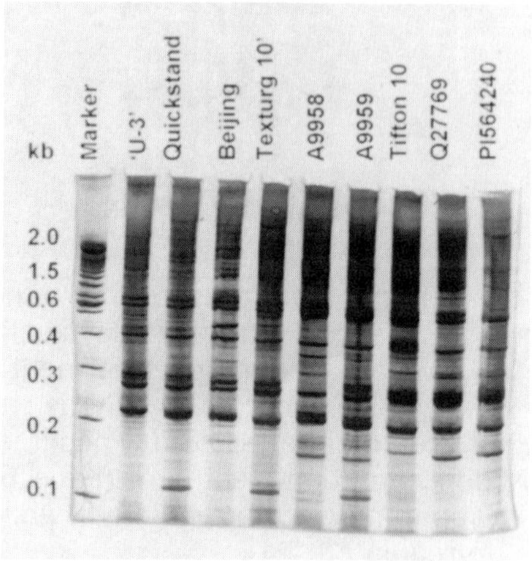

Figure 8.3. DAF gel using 1949 primer and genomic DNA extracts from nine *C. dactylon* genotypes.

in a previous DAF analysis (Caetano-Anollés et al., 1995). The farthest genetic distance was between Q27769 from Australia and U3, a bermudagrass cultivar from the United States. The geographic origin of Texturf 10 is the United States, but our analysis suggests some parentage from China, given the close relationship to the other Chinese bermudagrasses. Further work is in progress to expand the analysis to more lines of *C. dactylon* and to all nine *Cynodon* species using the Harlan and de Wet collection at Oklahoma State University. This will make it possible to compare the DAF PCR–based molecular approach with the previous taxonomic analysis based on morphology and other characteristics.

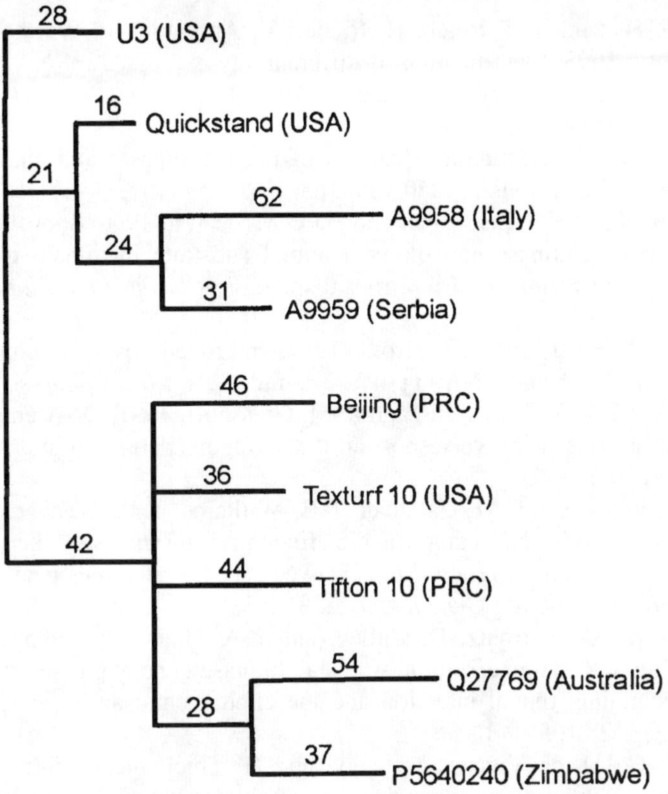

Figure 8.4. Phylogenetic relationships among nine bermudagrasses as determined by the DAF protocol, with national origin in parentheses. Numbers on branches represent branch length derived from PAUP 3.0 program.

REFERENCES

Anderson, J.A., M.P. Kenna, and C.M. Taliaferro. 1988. Cold hardiness of Midiron and Tifgreen bermudagrass. *HortScience*, 23:748–750.

Anderson, J.A., C.M. Taliaferro, and D.L. Martin. 1993. Evaluating freeze tolerance of bermudagrass in a controlled environment. *HortScience*, 28:955.

Antikainen, M. and S. Pihakaski. 1994. Early developments in RNA, protein, and sugar levels during cold stress in winter rye (*Secale cereale*) leaves. *Ann. Bot.*, 74:335–341.

Arlorio, M., A. Ludwig, T. Boller, and P. Bonfante. 1992. Inhibition of fungal growth by plant chitinases and beta-1,3-glucanases: a morphological study. *Protoplasma*, 171:34–43.

Bassam, B.J. and G. Caetano-Anollés. 1993. Silver staining of DNA in polyacrylamide gels. *Applied Biochemistry and Biotechnology*, 42:181–187.

Bauer, D., H. Muller, J. Reich, H. Riedel, V. Ahrenkiel, P. Warthoe, and M. Strauss. 1993. Identification of differentially expressed mRNA species by an improved display technique (DDRT-PCR). *Nucleic Acid Research*, 21:4272–4280.

Beintema, J.J. 1994. Structural features of plant chitinases and chitin binding proteins. *FEBS Letters*, 350:159–163.

Benhamou, N., M.H.A.J. Joosten, and P.J.G.M. de Wit. 1990. Subcellular localization of chitinase and of its potential substrate in tomato root tissues infected by *Fusarium oxysporum* f. sp. *radicis-lycopersici. Plant Physiol.*, 1108–1120.

Blum, H., H. Beier, and H.J. Gross. 1987. Improved silver staining of plant proteins, RNA and DNA in polyacrylamide gels. *Electrophoresis*, 8:93–99.

Caetano-Anollés, G., J. Bassam, and P. M. Gresshoff. 1991. DNA amplification fingerprinting using very short arbitrary oligonucleotide primers. *Biotechnology*, 9:553–557.

Caetano-Anollés, G., L.M. Callahan, P.E. Williams, K.R. Weaver, and P.M. Gresshoff. 1995. DNA amplification fingerprinting analysis of bermudagrass (*Cynodon*): genetic relationships between species and interspecific crosses. *Theoretical Applied Genetics*, 91:228–235.

Chang, M.M., D. Horovitz, D. Culley, and L.A. Hadwiger. 1995. Molecular cloning and characterization of a pea chitinase gene expressed in response to wounding, fungal infection and the elicitor chitosan. *Plant Molecular Biology*, 28:105–111.

Damerval, C., D. de Vienne, M. Zivy, and H. Thiellement. 1986. Technical improvements in two-dimensional electrophoresis increase the level of genetic variation detected in wheat seedling proteins. *Electrophoresis*, 7:52–54.

Danhash, N., C.A.M. Wagemakers, J.A.L. van Kan, and P.J.G.M. de Wit. 1993. Molecular characterization of four chitinase cDNAs obtained from Cladosporium fulvum–infected tomato. *Plant Molecular Biology*, 1017–1029.

de Wet, J.M.J. and J.R. Harlan. 1970. Biosystematics of *Cynodon* L.C. Rich. (*Gramineae*). *Taxon*, 19:565–569.

Farrell, R.E. 1993. *RNA Methodologies — A laboratory guide for isolation and characterization*. New York, Academic Press.

Fry, J.D. 1990. Cold temperature tolerance of bermudagrass. *Golf Course Management*, 58:26–32.

Fukuda, Y. and H. Shinshi. 1994. Characterization of a novel *cis*-acting element that is responsive to a fungal elicitor in the promoter of a tobacco class I chitinase gene. *Plant Molecular Biology*, 24:485–493.

Gatschet, M., C. Taliaferro, J. Anderson, and V. Baird. 1994a. Improving bermudagrass tolerance to winter stress. *Golf Course Management,* 62:52–55.

Gatschet, M.J., C.M. Taliaferro, J.A. Anderson, D.R. Porter, and M.P. Anderson. 1994b. Cold acclimation and alterations in protein synthesis in bermudagrass crowns. *Journal American Society Horticulture Science*, 119:477–480.

Gatschet, M.J. 1993. Alterations in protein synthesis associated with cold acclimation in bermudagrass (*Cynodon* spp.). Thesis, Department of Agronomy. Stillwater, Oklahoma State University.

Gatschet, M.J., C.M. Taliaferro, D.R. Porter, M.P. Anderson, J.A. Anderson, and K.W. Jackson. 1996. A cold regulated protein from bermudagrass crowns is a chitinase. *Crop Science*, 36:712–718.

Graham, L.S. and M.B. Sticklen. 1994. Plant chitinases. *Canadian Journal of Botany*, 72:1057–1083.

Griffith, M., P. Ala, D.S.C. Yang, W.C. Hon, and B.A. Moffatt. 1992. Antifreeze protein produced endogenously in winter rye leaves. *Plant Physiol.*, 100:593–596.

Guy, C.L. and D. Haskell. 1987. Induction of freezing tolerance in spinach is associated with the synthesis of cold acclimation induced proteins. *Plant Physiology*, 84:872–878.

Guy, C.L. and D. Haskell. 1988. Detection of polypeptides associated with the cold acclimation process in spinach. *Electrophoresis*, 9:787–796.

Harlan, J.R., J.M.J. de Wet, W.W. Huffine, and J.R. Deakin. 1970a. A guide to the species of *Cynodon* (*Gramineae*). *Oklahoma Agricultural Experiment Station Bulletin*, B-673.

Harlan, J.R., J.M.J. de Wet, and K.M. Rawal. 1970b. Geographic distribution of the species of *Cynodon* L.C. Rich (*Gramineae*). *Journal of East African Agriculture and Forestry*, 36:220–226.

Harlan, J.R., J.M.J. de Wet, K.M. Rawal, M.R. Felder, and W.L. Richardson. 1970. Cytogenetic studies in *Cynodon* L.C. Rich. (*Gramineae*). *Crop Science*, 10:288–291.

Hauschild, M.Z. 1993. Putrescine (1,4-diaminobutane) as an indicator of pollution-induced stress in higher plants: Barley and rape stressed with Cr(III) or Cr(VI). *Ecotoxicology and Environmental Safety*, 26:228–247.

Hausladen, A. and R.G. Alscher. 1994. Purification and characterization of glutathione reductase isozymes specific for the state of cold hardiness of red spruce. *Plant Physiol.*, 105:205–213.

Hincha, D.K., U. Heber, and J.M. Schmitt. 1990. Proteins from frost hardy leaves protect thylakoids against mechanical freeze-thaw damage *in vitro*. *Planta*, 180:416–419.

Hon, W.C., M. Griffith, A. Mlynarz, Y.C. Kwok, and D.S.C. Yang. 1995. Antifreeze proteins in winter rye are similar to pathogenesis related proteins. *Plant Physiology*, 109:879–889.

Johnson, R.R., H.J. Cranston, M.E. Chaverra, and W.E. Dyer. 1995. Characterization of cDNA clones for differentially expressed genes in embryos of dormant and nondormant *Avena fatua* L. caryopsis. *Plant Molecular Biology*, 28:113–122.

Kazuoka, T. and K. Oeda. 1994. Purification and characterization of COR85-oligomeric complex from cold-acclimated spinach. *Plant and Cell Physiol.*, 35:601–611.

Kim, Y.K., J.M. Baek, H.Y. Park, Y.D. Choi, and S.I. Kim. 1994. Isolation of characterization of cDNA clones encoding class I chitinase in suspension cultures of rice cell. *Bioscience Biotechnology and Biochemistry*, 58:1164–1166.

Kulsar, P. and G.D. Prestwich. 1988. Fluorography of tritium-labeled proteins in silver stained polyacrylamide gels. *Analytical Biochemistry*, 170:528–531.

Laemmli, U.K. 1970. Cleavage of structural proteins during assembly of the head of bacteriophage T4. *Nature*, 227:680–685.

Leborgne, N., C. Teulieres, S. Travert, M.P. Rols, J. Teissie, and A.M. Boudet. 1995. Introduction of specific carbohydrates into *Eucalyptus gunnii* cells increases their freezing tolerance. *Eur. J. Biochem.*, 229:710–717.

Liang, P., L. Averboukh, K. Keyomarsi, R. Sager, and A.B. Pardee. 1992. Differential display and cloning of messenger RNAs from human breast cancer versus mammary epithelial cells. *Cancer Research*, 52:6966–6968.

Liang, P., D. Bauer, L. Averboukh, P. Warthoe, M. Rohrwild, H. Muller, M. Strauss, and A.B. Pardee. 1995. Analysis of altered gene expression by differential display. *Methods in Enzymology*, 254:304–321.

Liang, P. and A.B. Pardee. 1992. Differential display of eukaryotic messenger RNA by means of the polymerase chain reaction. *Science*, 967–971.

Lin, C. and M.F. Thomashow. 1992. A cold-regulated Arabidopsis gene encodes a polypeptide having potent cryoprotective activity. *Biochem. Biophysics Research Communications*, 183:1103–1108.

Lin, W., C.S. Anuratha, K. Datta, I. Potrykus, S. Muthukrishnan, and S.K. Datta. 1995. Genetic engineering of rice for resistance to sheath blight. *Biotechnology*, 13:686–691.

Logemann, J., J. Schell, and L. Willmitzer. 1987. Improved method for the isolation of RNA from plant tissue. *Analytical. Biochemistry*, 163:16–20.

McCarty, L.B., L.T. Lucas, and J.M. DiPaola. 1990. Spring dead spot of bermudagrass. *Golf Course Management*, 58:36.

McKersie, B.D., Y. Chen, M. De Beus, S.R. Bowley, C. Bowler, D. Inze, K. D'Halluin, and J. Botterman. 1994. Superoxide dismutase enhances tolerance of freezing stress in transgenic alfalfa (*Medicago sativa* L.). *Plant Physiol.,* 103:1155–1163.

Mok, S.C., K.K. Wong, R.K.W. Chan, C.C. Lau, S.W. Tsao, R.C. Knapp, and R.S. Berkowitz. 1994. Molecular Cloning of Differentially Expressed Genes in Human Epithelial Ovarian Cancer. *Gynecologic Oncology*, 52:247–252.

Neuhaus, J.M., S. Flores, D. Keefe, P. Ahl Goy, and F. Meins, Jr. 1992. The function of vacuolar beta-1,3-glucanase investigated by antisense transformation. Susceptibility of transgenic *Nicotiana sylvestris* plants to *Cercospora nicotianae* infection. *Plant Mololecular Biology,* 19:803–813.

Payne, G., P. Ahl, M. Moyer, A. Harper, J. Beck, F. Meins, and J. Ryals. 1990. Isolation of complementary DNA clones encoding pathogenesis related proteins P and Q, two acidic chitinases from tobacco. *Proceedings National Academy of Science*, 87:98–102.

Perras, M. and F. Sarhan. 1989. Synthesis of freezing tolerance proteins in leaves, crown, and roots during cold acclimation of wheat. *Plant Physiol.*, 89:577–585.

Porter, D.R. and M.J. Gatschet. 1992. Simplified drying of polyacrylamide gels for fluorography. *Biotechniques*, 13:364–365.

Porter, D.R., J.T. Weeks, M.P. Anderson, and A.C. Guenzi. 1992. An easy technique for extruding polyacrylamide gels from isoelectric focusing tubes of 1.0 to 1.5 mm inside diameter. *Biotechniques*, 12:380–382.

Rasmussen, U., H. Giese, and J.D. Mikkelsen. 1992. Induction and purification of chitinase in *Brassica napus* L. spp. *oleifera* infected with *Phoma lingam. Planta,* 187:328–334.

Roby, D., K. Broglie, J. Gaynor, and R. Broglie. 1991. Regulation of a chitinase gene promoter by ethylene and elicitors in bean protoplasts. *Plant Physiol.*, 97:433–439.

Rochester, D.E., A. Brands, Y. Zhu, C. Myers, and T.H.D. Ho. 1995. Use of the differential display technique to clone early gibberellic acid induced cDNAs from barley aleurone tissue. *Plant Physiology*, 108:80.

Rosenfeld, J., J. Capdevielle, J.C. Guillemot, and P. Ferrara. 1992. In-gel digestion of proteins for internal sequence analysis after one- or two-dimensional gel electrophoresis. *Analytical Biochemistry*, 203:173–179.

Sager, R., A. Anisowicz, M. Neveu, P. Liang, and G. Sotiropoulou. 1993. Identification by differential display of alpha 6 integrin as a candidate tumor suppressor gene. *FASEB Journal,* 7:964–970.

Sakai, A. and W. Larcher. 1987. *Frost Survival of Plants*. Berlin, Springer-Verlag.

Sambrook, J., E.F. Fritsch, and T. Maniatis. 1989. *Molecular cloning — A laboratory manual*, Cold Spring Harbor Laboratory Press.

Sastry, J.G., W. Ramakrishna, S. Sivaramakrishnan, R.P. Thakur, V.S. Gupta, and P.K. Ranjekar. 1995. DNA fingerprinting detects genetic variability in pearl millet downy mildew pathogen (*Sclerospora graminicola*). *Theoretical Applied Genetics*, 91:856–861.

Sharma, Y.K. and K.R. Davis. 1995. Isolation of a novel arabidopsis ozone-induced cDNA by differential display. *Plant Molecular Biology*, 29:91–98.

Shinshi, H., D. Mohnen, and F. Meins, Jr. 1987. Regulation of a plant pathogenesis-related enzyme: inhibition of chitinase and chitinase mRNA accumulation in cultured tobacco tissues by auxin and cytokinin. *Proc. Nat. Acad. Sci.*, 84:89–93.

Sieg, F., W. Schroder, J.M. Schmitt, and D.K. Hincha. 1996. Purification and characterization of a cryoprotective protein (cryoprotectin) from leaves of cold acclimated cabbage. *Plant Physiology*, 111:215–221.

Smiley, R.W, P.H. Dernoeden, and B.B. Clarke. 1992. *Compendium of Turfgrass Diseases*. 2nd Ed. APS Press, St. Paul, MN.

Song, P., E. Yamamoto, R.P. Webb, and R.D. Allen. 1995. Cloning of full length cotton fiber specific cDNA's by differential display and PCR library screening. *Plant Physiology*, 108 suppl., 134.

Steponkus, P.L. 1984. Role of the plasma membrane in freezing injury and cold acclimation in plants. *Annual Review of Plant Physiology*, 35:543–584.

Sutton, F., K. Han, and D.G. Kenefick. 1995. Differential display reveals genes related to freeze resistance in wheat. *Plant Physiology*, 108 Suppl., 153.

Swofford, D.L. 1993. PAUP: Phytogenetic Analysis Using Parsimony, Version 3.0, formerly distributed by Illinois Natural History Survey, Champaign, IL, but soon available through Sinauer Associates, Sunderland, MA as Version 4.0.

Taliaferro, C.M. 1995. Diversity and vulnerability of bermuda turfgrass species. *Crop Science*, 35:327–332.

Thomashow, M.F. 1990. Molecular genetics of cold acclimation in higher plants. *Advances in Genetics*, 28, 99–131.

Tieman, D.M. and A.K. Handa. 1995. Molecular cloning and characterization of genes expressed during early tomato (*Lycopersicon esculentum* Mill.) fruit development by mRNA differential display. *Journal of the American Society for Horticultural Science,* 121:52–56.

Tronsmo, A.M., P. Gregersen, L. Hjeljord, T. Sandal, T. Bryngelsson, and D.B. Collinge. 1993. Cold Induced Disease Resistance. *Mechanisms of Plant Defense Response*. B. Fritig and M. Legrand. Kluwer Academic Publishers, Netherlands.

Tseng, T.C., T.H. Tsai, M.Y. Lue, and H.T. Lee. 1995. Identification of sucrose regulated genes in cultured rice cells using mRNA differential display. *Gene*, 161/2:179–182.

Uemura, M., R.A. Joseph, and P.L. Steponkus. 1995. Cold acclimation of arabidopsis: Effect on plasma membrane lipid composition and freeze induced lesions. *Plant Physiology*, 109:15–30.

Watson, M.A. and T.P. Fleming. 1994. Isolation of differentially expressed sequence tags from human breast cancer. *Cancer Research*, 54:4598–4602.

Webb, M.S. and P.L. Steponkus. 1993. Freeze induced membrane ultrastructural alterations in rye (*Secale cereale*) leaves. *Plant Physiology*, 101:955–963.

Webb, M.S., M. Uemura, and P.L. Steponkus. 1994. A comparison of freezing injury in oat and rye: two cereals at the extremes of freezing tolerance. *Plant Physiol.*, 104:467–478.

Wolfraim, L.A., R. Langis, H. Tyson, and R.S. Dhindsa. 1993. cDNA sequence, expression and transcript stability of a cold acclimation-specific gene, cas18, of alfalfa (*Medicago falcata*) cells. *Plant Physiology*, 101:1275–1282.

Yamagami, T. and G. Funatsu. 1993. The complete amino acid sequence of chitinase-c from the seeds of rye (*Secale cereale*). *Bioscience Biotechnology and Biochemistry*, 57:1854–1861.

Chapter 9

Alterations of Membrane Composition and Gene Expression in Bermudagrass During Acclimation to Low Temperature

Wm.V. Baird, S. Samala, G.L. Powell, M.B. Riley, J. Yan, and J. Wells

INTRODUCTION

Environmental stress occurs in many forms, such as low temperature or water deficit. The physical and biochemical changes that occur in chilling sensitive plants exposed to reduced temperature, together with the subsequent expression of low-temperature stress symptoms, are collectively referred to as cold or chilling injury. The physiological changes include alterations in cytoplasmic streaming, enzyme activity, respiration, and photosynthesis, as well as effects on membrane permeability, structure, and composition. Which if any of these changes is responsible for the primary low temperature–induced injury remains uncertain.

It is clear, however, that disruption of cellular membrane integrity, as a result of low temperature–induced water removal, is a primary cause of cold injury. The membrane lipid bilayer provides the necessary environment for proper functioning of proteins and enzymes associated with a particular membrane system. Maintenance of a fluid state for membrane lipids is thought to be one of the prerequisites for unimpaired survival at low temperature.

Physical properties such as flexibility and molecular motions of membrane diacylglycerols (e.g., polar phosphoglycerides) depend to a large extent on the degree of unsaturation in their fatty acid (FA) side chains (Quinn et al., 1985; Wada et al., 1994). Thus, changes in FA unsaturation can affect biophysical properties such as the temperature at which membranes undergo dehydration-induced phase transitions (Lyons, 1973; Palta et al., 1993).

Weiser (1970), working primarily with woody plant species, was the first to propose that cold acclimation and low temperature survival involve fundamental changes in the physiology of the plant at the level of gene expression. This concept was extended in studies of cold acclimation in spinach (Guy et al., 1985). Mohapatra and co-workers (1989) obtained very high correlation coefficients between levels of expression of cold-induced genes and the degree of freezing tolerance in acclimated alfalfa. Since then, several workers have documented a causal relationship between exposure of plants to low temperature and increases in gene expression or induction of previously unexpressed genes (e.g., Crespi et al., 1991; Hajela et al., 1990; Lin et al., 1990; Nordin et al., 1991; Perras and Sarhan, 1989; Thomashow et al., 1990; Vallejos, 1991; Weiser et al., 1990). Whether these genes are biologically significant (Guy, 1990a,b; Thomashow, 1990) in conferring tolerance to the effects of low temperature, e.g., code for enzymes involved in lipid or FA metabolism, and whether they have homologs in bermudagrass or other turf species, remains unknown at this time. One notable exception is the recent identification of a category b-II chitinase, with presumed antifungal and/or cryoprotective properties, expressed in crowns of bermudagrass in response to cold (Gatschet et al., 1996).

A better understanding of the nature and basis for tolerance to low temperature and the process of cold acclimation in bermudagrass would be very helpful for programs focused on germplasm improvement through breeding or biotechnology. In the preliminary studies reported on here, we investigated the effect of low temperature (i) on the FA composition of total polar lipids, and (ii) on gene expression during cold acclimation.

MATERIALS AND METHODS

Plant Material and Growth Conditions

The bermudagrass cultivars used in this study were Midiron, Tifway, Tifgreen, and U3. The plants were grown and vegetatively propagated in soilless medium in "Rootrainers," and maintained in the greenhouse with an automated watering system. For cold acclimation experiments, preconditioning of plants was for 2 weeks in controlled environment chambers under conditions essentially as outlined previously (Anderson et al., 1993) — 10 hr photoperiod, 250 μm/m^2/s PPFD, 29/24°C (d/n). Cold acclimation was similarly achieved as previously described (Anderson et al., 1988; 1993) — 10 hr photoperiod, 250 μm/m^2/s, 8/3°C (d/n). These experiments were performed for 2–4 weeks in controlled environment chambers.

Membrane Lipid Isolation and Analysis

Membrane lipids were isolated according to the procedure of Bligh and Dyer (1959). Total lipid extractions were fractionated into neutral lipids, glycolipids, and phospholipids by sequential organic elution from silica-based chromatography columns (Lynch and Steponkus, 1987). Separation of individual lipids was accomplished by thin layer chromatography (TLC), and identified by cochromatography with authentic standards and the use of spray reagents (Kates, 1972).

Fatty Acid Analysis

Lipids were saponified to release their FA components, and then these were converted to methyl esters. The mixture of FA methyl esters was analyzed by gas chromatography, and the individual species were identified by their retention times using a computer-based identification system (MIDI, Inc., Newark, Delaware). The amounts of each species were quantified by comparing the area of their peak with that of a known internal standard. Any ambiguous peaks were resolved using mass spectroscopy and/or authentic standards.

Nucleic Acid Isolation

Genomic DNA was isolated by standard mechanical and chemical methods that employ cell wall disruption, membrane lysis, nuclease inhibition, and protein, carbohydrate, and lipid removal prior to nucleic acid concentration (Sambrook et al., 1989; Ausubel et al., 1994). Southern hybridization blots were prepared, basically as described in Sambrook et al. (1989), using restriction enzymes according to manufacturer's specifications, and probed with radiolabeled plasmid inserts of specific gene sequences. Total RNA was isolated using a standard salt (LiCl) precipitation method, and further treated with DNAse prior to phenol/chloroform extraction.

Analysis of Gene Expression

Messenger RNA profiling gels ("differential display," DD/RT-PCR) were prepared using, as a substrate, total RNA isolated at different time points following induction of cold acclimation. The methods followed published protocols of Liang and Pardee (1992) and Bauer et al. (1993). Briefly, anchor primers, either specific (i.e., $dT_{11}NN$ or $dT_{12}VN$; where V = dA, dG and dC and N = dA, dG, dC or dT), were used for first strand cDNA synthesis by reverse transcriptase. The single-strand cDNAs were amplified by PCR using an arbitrarily chosen (downstream) decamer primer and the same anchor primer. ^{32}Phosphorus-labeled reactions from treated and nontreated tissue were fractionated on nondenaturing polyacrylamide gels. Individual bands exhibiting differential expression between treated and nontreated samples were identified for reamplification, cloning, and further characterization.

RESULTS AND DISCUSSION

Alteration in Membrane Fatty Acid Composition

Our preliminary findings indicated that different organs of the same genotype responded differentially to low temperature. For example, crowns showed the largest sustained increase in FA content (per mg of total lipid) over the four-week experiment, especially when contrasted to leaf tissues. Even the response in roots was attenuated as compared to crowns. This trend was the same for the two cultivars examined, although the magnitude of the change for U3 (e.g., relatively cold-sensitive) was not as great as for 'Midiron' (e.g., relatively cold-tolerant).

Overall, significantly greater than 90% of the total polar FA content was accounted for by less than a half dozen FA species. Some of the most abundant species appeared to be: palmitic acid (16:0 = 16 carbon chain-length and no double bonds), stearic acid (18:0), linoleic acid (18:2) and linolenic acid (18:3). 'Midiron' responded more rapidly and to a greater extent than did U3 for the changes in FA composition reported in this study. This was illustrated by the greater than three-fold increase of unsaturated FA to saturated FA ratio for 'Midiron' over U3. Similarly, the difference between the double bond index (DBI) for the two genotypes is also indicative of this trend. In addition, although the DBI was essentially identical for the two genotypes at the start of the experiments, the observed increase in DBI during cold acclimation was initiated earlier and increased more rapidly in 'Midiron' than it did in U3.

The observed alterations in membrane FA content presented in this preliminary report (e.g., increased levels of 18:3, and decreased levels of 16:0) are consistent with studies in other species. These show correlations of increases in membrane lipid FA unsaturation with enhanced cold tolerance. They emphasize the importance of FA unsaturation and its role in membrane fluidity — thus protecting membranes from undergoing damaging phase transitions at higher temperatures. In addition, our results point to specific desaturase (e.g., ω-3 and/or Δ-12) and transferase-type (e.g., 3-keto-acyl-ACP synthase

II) enzymes as being of primary importance in regulating membrane lipid FA composition in response to low temperature, and ultimately in avoiding the winter damage suffered by bermudagrass along the warm-season/cool-season transition zone.

The measured increase in 18:3 is most likely the consequence of the activity of a linoleic acid (ω-3) desaturase. Because this increase occurred in response to low temperature, the gene for this (iso)enzyme may be temperature-regulated, especially so in 'Midiron'. Our findings point to the need for more detailed and in-depth investigations of the lipids and their constituent FAs from specific membrane-enriched fractions of bermudagrass, as well as employing other cultivars, differing in their ability to cold acclimate, in these investigations. Additionally, molecular studies of the genes for the lipid and FA biosynthesis enzymes are now possible because of the availability of heterologous probes for a number of these genes from various higher plant species.

Stress-Regulated Gene Homologies in Bermudagrass

All four cultivars of bermudagrass (Midiron, U3, Tifgreen, and Tifway) were used as experimental material in screens for the presence of low-temperature and/or drought-induced gene sequences in the bermudagrass nuclear genome. Southern blot membranes, containing restriction endonuclease digested total genomic DNA from each cultivar, were prepared and screened with heterologous gene probes. Twelve plasmids containing genomic or cDNA clones, expressed during exposure to low temperature or water deficit in other plant systems (i.e., *Arabidopsis*, barley, maize, spinach, tobacco, tomato, and wheat), were used as gene probes. Because of the heterologous nature of the probes, molecular hybridizations were performed at various levels of stringency to screen for related sequences in the bermudagrass genome, while avoiding spurious cross-hybridization signals.

Four categories of hybridization signals were observed. One class detected homologous gene sequences (more than one fragment), and the overall banding patterns showed variation between the four bermudagrass cultivars (i.e., restriction fragment length polymorphisms = RFLPs). These probes may be useful in comparisons between germplasm (e.g., exotic accessions, hybrids, segregating populations, etc.) for gross evaluations of the genetic variation associated with their potential response to low temperatures and/or drought conditions. One example of a clone in this class is a cold-regulated gene sequence from *Arabidopsis*, which hybridizes to between 5 and 7 fragments (i.e., gene family) in the bermudagrass genome, depending upon the cultivar and restriction enzyme used.

Similarly, a second group detected complementary sequences in the bermudagrass genome, but showed little or no variation between the four cultivars examined. Some of the members in this class detected ten or more "homologous" sequences, occasionally revealing minor hybridization pattern differences (i.e., low signal intensities) between the genotypes. These gene clones represented cold-regulated sequences from *Arabidopsis* and wheat. One member, a drought-induced gene sequence from tobacco, detected only a single fragment, and it was monomorphic (no variation) in all four genotypes.

The third grouping (represented by only a single member whose transcript accumulates in *Arabidopsis* in response to low temperature) was interesting in that it showed a very strong hybridization signal to only a single sequence in a single cultivar (i.e., U3). The biological significance of this finding is unknown. It may be the result of cross-hybridization to an unrelated but highly abundant sequence or, more interestingly, it may be related to a physiological capability unique to the U3 genotype.

A fourth group of clones showed no significant homology to nuclear DNA sequences in bermudagrass, even at low hybridization stringency. Interestingly, many of the clones that compose this class (e.g., three from tomato and one from corn) are induced by drought/dehydration or by the exogenous application of the plant hormone abscisic acid. These heterologous gene clones were deemed uninformative, and were not used in future analyses.

Differential Gene Expression

Differential display/RNA profiling has been used to identify genes expressed in bermudagrass during cold acclimation. To date, over 60 variable-anchor:10mer or specific-anchor:10mer primer pair combinations have been screened. Of these, approximately 17 have exhibited possible differential expression with one or more bands.

As an example, four up-regulated sequences, putatively expressed in response to low temperature, were identified through comparisons of the gel-based "displays" from 0, 1, 2, and 7 days post–low temperature exposure to similarly derived "displays" from control, nonacclimating tissue over the same time frame, from Midiron and U3. Expression of these sequences appears to be induced in response to low temperature treatment, as indicated by their appearance at 24 hr and continued presence during longer exposures, as well as their complete absence from reactions using RNA isolated from time "zero" and nonacclimating control tissues. These sequences ranged from 300 to 600 base pairs in size.

Such fragments, following second round PCR amplification using the same primer pairs, have been blunt-end cloned into multicopy plasmids for further characterization. These experiments will include DNA sequence determination for homology searches of gene databases, as well as their use as probes on Northern blots, or in RNA protection assays, to confirm their differential, cold-specific expression.

ACKNOWLEDGMENTS

Support for this work was provided by a grant from the U.S. Golf Association, Green Section Research, and by the South Carolina Agriculture Experiment Station, Clemson University.

REFERENCES

Anderson, J.A., M.P. Kenna, and C.M. Taliaferro. 1988. Cold hardiness of 'Midiron' and 'Tifgreen' bermudagrass. *HortScience,* 23:748–750.

Anderson, J.A., C.M. Taliaferro, and D.L. Martin. 1993. Evaluating freeze tolerance of bermudagrass in a controlled environment. *HortScience*, 28:955.

Ausubel, F., R. Brent, R. Kingston, D. Moore, J. Seidman, J. Smith, and K. Struhl. 1994. *Current Protocols in Molecular Biology*. John Wiley & Sons, NY.

Bauer, D., H. Muller, J. Reich, H. Riedel, V. Ahrenkiel, P. Warthoe, and M. Strauss. 1993. Identification of differentially expressed mRNA species by an improved display technique (DDRT-PCR). *Nuc. Acids Res.,* 2118:4272–4280.

Bligh, E.G. and W.J. Dyer. 1959. A rapid method of total lipid extraction and purification. *Can J. Biochem. Physiol.,* 37:911–917.

Close, T., A. Kortt, and P. Chandler. 1989. A cDNA-based comparison of dehydration-induced proteins (dehydrins) in barley and corn. *Plant Mol. Biol.,* 13: 95–108.

Crespi, M., E. Zabaleta, H. Pontis, and G. Salerno. 1991. Sucrose synthase expression during cold acclimation in wheat. *Plant Physiol.,* 96:887–891.

Gatschet, M., C.M. Taliaferro, D.R. Porter, M.P. Anderson, J.A. Anderson, and K.W. Jackson. 1996. A cold-regulated protein from bermudagrass crowns is a chitinase. *Crop Sci.,* 36:712–718.

Guy, S., K. Niemi, and R. Bramble. 1985. Altered gene expression during cold acclimation of spinach. *Proc. Nat. Acad. Sci., USA,* 82:3673–3677.

Guy, C. 1990a. Cold acclimation and freezing stress tolerance: Role of protein metabolism. *Ann. Rev. Plant Physiol. Mol. Biol.,* 41:187–223.

Guy, C. 1990b. Molecular mechanisms of cold acclimation. pp. 35–61, in *Environmental Injury to Plants.* F. Ketterman (Ed.). Academic Press, NY, p. 290.

Hajela, R., D. Horvath, S. Gilmour, and M.F. Thomashow. 1990. Molecular cloning and expression of *cor* (*co*ld-*r*egulated) genes in *Arabidopsis thaliana. Plant Physiol.* 93:1246–1252.

Kates, M. 1972. Techniques of lipidology, in *Laboratory Techniques in Biochemistry and Molecular Biology*, 3rd ed. Work, T.S. and E. Work (Eds.). Elsevier/North-Holland Press, Amsterdam, p. 464.

Liang, P. and A.B. Pardee. 1992. Differential display of eukaryotic messenger RNA by means of the polymerase chain reaction. *Science*, 257:967–971.

Lin, C., W. Guo, E. Everson, and M.F. Thomashow. 1990. Cold acclimation in *Arabidopsis* and wheat: A response associated with expression of related genes encoding 'boiling-stable' polypeptides. *Plant Physiol.,* 94:1078–1083.

Lynch, D.V. and P.L. Steponkus. 1987. Plasma membrane lipid alterations associated with cold acclimation of winter rye seedlings (*Secale cereale* L. cv. Puma). *Plant Physiol.,* 83:761–767.

Lyons, J.M. Chilling injury in plants. 1973. *Ann. Rev. Plant Physiol.* 24:445–466.

Mohapatra, S., L. Wolfraim, R. Poole, and R. Dhindsa. 1989. Molecular cloning and relation to freezing tolerance of cold-acclimation-specific genes in alfalfa. *Plant Physiol.* 89:375–380.

Nordin, K., P. Heino, and E.T. Palva. 1991. Separate signal pathways regulate the expression of a low-temperature-induced gene in *Arabidopsis thaliana* (L.) Heynh. *Plant Mol. Biol.,* 16:1061–1071.

Palta, J.P., B.D. Whitaker, and L.S. Weiss. 1993. Plasma membrane lipids associated with genetic variability in freezing tolerance and cold acclimation of *Solanum* species. *Plant Physiol.*, 103:793–803.

Perras, M. and F. Sarhan. 1989. Synthesis of freezing tolerance proteins in leaves, crowns and roots during cold acclimation of wheat. *Plant Physiol.,* 89:577–585.

Quinn, P.J., F. Joo, and L. Vigh. 1985. The role of unsaturated lipids in membrane structure and stability. *Prog. Biophys. Mol. Biol.,* 53:71–103.

Sambrook, J., E.F. Fritsch, and T. Maniatis. 1989. *Molecular Cloning: A laboratory manual.* 2nd Ed. Cold Spring Harbor, NY: Cold Spring Harbor Laboratory Press.

Thomashow, M.F. 1990. Molecular genetics of cold acclimation in higher plants. pp. 99–131, in *Advances in Genetics.* Scandalios, J.G. (Ed.). Vol. 28. Academic Press, NY. p. 308.

Thomashow, M.F, S. Gilmour, R. Hajela, D. Horvath, C. Lin, and W. Guo. 1990. Studies on cold acclimation in *Arabidopsis thaliana*. pp. 305–314, in *Horticultural Biotechnology.* Bennett, A. (Ed.). Wiley-Liss, Inc. NY.

Vallejos, E. 1991. Low night temperatures have a differential effect on the diurnal cycling of gene expression in cold-sensitive and tolerant tomatoes. *Plant, Cell Environ.,* 14:105–112.

Wada, H., Z. Gombos, and N. Murata. 1994. Contribution of membrane lipids to the ability of the photosynthetic machinery to tolerate temperature stress. *Proc. Nat. Acad. Sci. USA,* 91:4273–4277.

Weiser, C.J. 1970. Cold resistance and injury in woody plants. *Science,* 169:1269–1278.

Weiser, R.L., S. Wallner, and J. Waddell. 1990. Cell wall and extension mRNA changes during cold acclimation of pea seedlings. *Plant Physiol.,* 93:1021–1026.

Chapter 10

Analysis of Heat Shock Response in Perennial Ryegrass

P.M. Sweeney, T.K. Danneberger, J.A. DiMascio, and J.C. Kamalay

The accurate evaluation of thermal tolerance in perennial ryegrass (*Lolim perenne* L.) is important in cultivar development and selection. In temperate regions, high temperature stress is a contributing factor to the decline of perennial ryegrass. During periods of high temperature, thermal-tolerant perennial ryegrass cultivars are more competitive and hence more desirable. Alternately, in subtropical regions, heat-sensitive perennial ryegrass may be preferred when overseeding dormant warm-season species. A more competitive thermal-tolerant perennial ryegrass may hinder the growth and development of the warm-season turfgrass as it breaks dormancy during higher temperatures in the spring.

Assessment of thermal tolerance based on field performance is complicated by the difficulty of separating plant response to increased temperature from other environmental factors, particularly water stress. Measurement of electrolyte leakage has been proposed as a way to quantify thermal tolerance differences (Sullivan and Ross, 1979). Wallner et al. (1982) and White et al. (1988) found electrolyte leakage of perennial ryegrass leaf segments subjected to high temperatures (55°C) to be greater than that of tissue subjected to lower temperatures (52° and 51°C). Wallner et al. (1982) reported that electrolyte leakage from leaf segments subjected to high temperatures was lower when the segments were acclimated to a high but nonlethal temperature (40°C), for 7 days prior to the heat treatment.

Results by Wallner et al. (1982) suggest that heat shock proteins (HSP) may be involved in the thermal tolerance of perennial ryegrass. Heat shock response (HSR) occurs in a number of species, including maize (*Zea mays* L.) (Baszcynski et al., 1982; Cooper and Ho, 1983) and wheat (*Triticum aestivum* L.) (Hendershot et al., 1992). The HSR is characterized by the induction of a set of thermally induced mRNA and their translation products: HSPs. During periods of high temperature stress, the synthesis of normal cellular proteins may be reduced or stopped, while HSP synthesis is initiated. These HSPs consist of a set of high-molecular-weight polypeptides (68, 70, 84, and 90 kilodalton (kD)) and a more prominent and complex set of low-molecular-weight polypeptides (15–17 kD) (Kimpel and Key, 1985; Vierling, 1991). Although the exact function of HSPs is not fully understood, their presence has been associated with the development of thermal tolerance (Vierling and Nguyen, 1992). Some research suggests that HSPs act as chaperonins in folding and assembling proteins in the post-translation stage of synthesis (Beckmann et al., 1990).

If HSPs are responsible for some portion of thermal tolerance differences exhibited by perennial ryegrass cultivars, it may be possible to test turfgrass cultivars and genotypes for thermal tolerance by evaluating heat-induced gene products or polymorphic thermal tolerance genes. The objective of the study was to evaluate the possibility of using HSP genes or gene products to screen for thermal tolerance in perennial ryegrass. Specifically, we wished to evaluate differences in HSP in two perennial ryegrass cultivars of differing thermal tolerance.

MATERIALS AND METHODS

Two perennial ryegrass cultivars, "Accolade," a thermal-tolerant cultivar (O.M. Scott and Sons Company, 1990) and "Caravelle" a thermal-sensitive cultivar (Mommersteeg International B.V., 1980), were used in the experiments. For each, 50 seeds of each cultivar were placed on 8 g/L water agar and incubated at 28°C for 4 days. Seedlings were then subjected to the following treatments: 40°C for 1 hr, 40°C for 1 hr followed by 55°C for 2 hr, or 55°C for 2 hr on day 4 (Table 10.1). Controls that were continuously at 28°C were included. Following treatments, all seedlings were placed at 28°C. On day 7, measurements of root, shoot, and total seedling length were made. Average seedling measurements from three replications of each treatment were used in paired *t*-test for analysis of the data. Genomic DNA was isolated from bulked samples of whole seedlings as described by Baker, et al. (1990), digested with *Eco*R I and *Hind* III, blotted onto nylon membranes, and hybridized to two maize HS clones [70 kD HSP (pMON9501; Rochester et al., 1986) and maize 26 kD HSP (pZmHSP26; Nieto-Sotelo et al., 1990)] that had been labeled with P^{32}. Total RNA was isolated from seedlings immediately following thermal treatment in the same manner. After formaldehyde-agarose gel separation

Table 10.1. Overview of the Heat Stress Treatments to Perennial Ryegrass Seedlings.

1 to 4 days	4 days	4 to 7 days	7 days
28°C	28°C; RNA extracted	28°C	Growth parameters measured
28°C	40°C 1 hr; RNA extracted	28°C	Growth parameters measured
28°C	40°C 1 hr; 55°C 2 hr; RNA extracted	28°C	Growth parameters measured
28°C	55°C 2 hr; RNA extracted	28°C	Growth parameters measured

and transfer to nylon membranes, the RNA was hybridized to the same maize HS clones.

Total RNA was translated using a cell free wheat germ extract system in the presence of ^{35}S-methionine. Incorporation was determined by trichloroacetic acid precipitations and the counts were monitored. After adjusting loads to ensure that a similar number of counts was added to each lane, translation products were separated by one-dimensional sodium dodecyl sulfate polyacrylamide gel electrophoresis (DiMascio et al., 1994).

RESULTS AND DISCUSSION

Under non-stress conditions (28°C), there was no significant difference (P=0.05) between "Accolade" and "Caravelle" for root, shoot, or total seedling length (Table 10.2). After exposure to 55°C for 2 hr, all seedlings died. Seedlings that received the 40°C acclimation treatment were not killed at 55°C.

After heat treatment, growth differences were detected between "Accolade" and "Caravelle" seedlings. "Accolade" shoot and total seedling length were significantly (P=0.05) greater than those of "Caravelle" after 1 hr at 40°C (Table 10.2). "Accolade" also had significantly (P=0.05) greater root, shoot, and total seedling length when the 40°C acclimation preceded the 2 hr 55°C treatment and after the 55°C treatment (Table 10.2). The increased growth of "Accolade" relative to "Caravelle," after exposure to 40°C or 40°C followed by 55°C, may indicate that the response of "Accolade" seedlings to heat shock is different than "Caravelle" seedlings. Both cultivars are synthetic varieties and are populations rather than single genotypes. The difference between the cultivars' response to acclimation may be due to differences in the proportion of individuals in each that exhibit acquired thermal tolerance rather than qualitative differences in heat shock response at the molecular level for each individual in the respective populations.

Table 10.2. Average Root, Shoot, and Total Length of 7-Day-Old "Accolade" (A) and "Caravelle" (C) Seedlings Incubated at 28°C and Subjected to Varying Heat Treatments on Day 4 of the Incubation.

	Length of Growth Parameters (cm)					
	Root		Shoot		Total Seedling	
	A	C	A	C	A	C
28°C	4.32	3.43	5.29	4.18	9.61	7.62
40°C: 1 hr	3.35	2.72	4.30[a]	2.65	7.65[a]	5.37
40°C: 1 hr, 55°C: 2 hr	3.29[a]	2.27	3.74[a]	2.45	7.03[a]	4.72
55°C: 2 hr	1.75[a]	1.26	1.66[a]	0.93	3.41[a]	2.19

[a] Significant at the 0.05 probability level for *t*-tests between A and C cultivars.

A 75 kD peptide unique to translation products of RNA extracted from seedlings of both "Accolade" and "Caravelle" after the 40°C heat treatment was observed (Figure 10.1). Although mRNA was present in the seedlings that experienced the 55°C treatment without prior acclimation (Figure 10.2), *in vitro* translation indicated that few were translatable (Figure 10.1). Increases in relative amounts of translatable mRNA were observed at approximately 14, 32 (40°C) and 25 kD (40/55°C) in the *in vitro* translation products from RNA extracts from seedlings that experienced acclimation (Figure 10.1). The patterns of the polypeptides resulting from *in vitro* translation of mRNA were similar for "Accolade" and "Caravelle." No qualitative differences were observed (Figure 10.1). The observation of new mRNA in the seedlings that experienced acclimation temperatures suggests that acclimation of perennial ryegrass to lethal temperatures was accompanied by an alteration of gene expression and supports the hypothesis that HSPs are involved in the acclimation of perennial ryegrass to lethal temperatures.

The heat-tolerant and heat-sensitive cultivar shared many DNA fragments detected by the maize HSP clones (Figure 10.3). There was, however, some indication that the maize clones could be used to reveal sequence polymorphisms. Restriction fragment length polymorphisms of 3.0, 5.1, and 5.3 kb were detected by the hybridization of HSP 70 to the *Hin*d III restrictions (Figure 10.3).

Northern blot hybridizations to the maize HSP-70 clone revealed a major heat-inducible transcript of 2.2 kb, and two minor transcripts at 2.8 and 1.1 kb (Figure 10.2A) in both cultivars. Much higher levels of the 2.2 kb messages were seen in the RNA from seedlings that experienced 1 hr 40°C or 2 hr 55°C heat stress (Figure 10.2A). Although the level of the minor HSP-70-like transcripts showed less change following heat treatments than did the 2.2 kb transcript, levels of all three mRNA appeared to be increased by the elevated incubation temperatures.

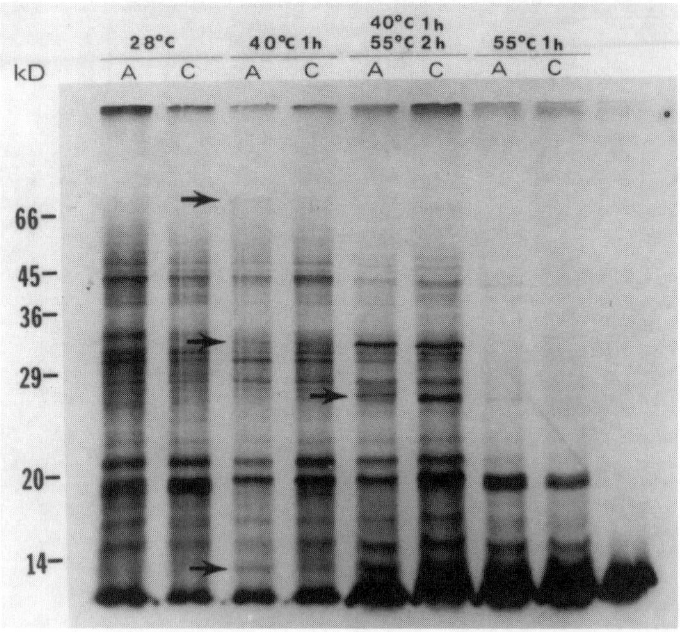

Figure 10.1. *In vitro* translation products of total RNA from "Accolade" (A) and "Caravelle" (C) perennial ryegrass seedlings subjected to 40°C for 1 hr; 40°C for 1 hr followed by 2 hr at 55°C; and 55°C for 2 hr on day 4 of incubation. Control (28°C) was not subjected to heat treatment. Fluorograph represents sodium dodecyl sulfate polyacrylamide gel electrophoresis of *in vitro* translation products stimulated by the addition of perennial ryegrass total RNA (10 μg per reaction) to wheat germ lysates. Approximately 100,000 trichloroacetic acid precipitable counts from each reaction were loaded per lane.

We were surprised to observe a reduction in the 2.2 kb HSP-70 transcript when seedlings were exposed to the 40°C/55°C treatment (Figure 10.2A). Exposure of seedlings to either the 40°C or 55°C temperature alone resulted in elevated levels of the 2.2 kb transcript. It appears that at the time of RNA extraction, which, for the 40°C/55°C treatment followed 3 hr of exposure to elevated temperatures, the heat-induced HSP-70 mRNAs were degraded. Although the timing of mRNA synthesis and degradation varies among HSPs, there is evidence that heat-induced mRNAs are transient (Kimpel and Key, 1985; DeRocher et al., 1991).

Maximum levels of a 1.1 kb HSP-26 transcript were detected in both cultivars after incubation for 1 hr at 40°C (Figure 10.2B). Hybridizable mRNA levels were only slightly lower after the 40°C/55°C treatment than after the 40°C treatment. The 1.1 kb HSP-26 message was not detectable in the RNA extracted from the seedlings that were incubated at 55°C without prior accli-

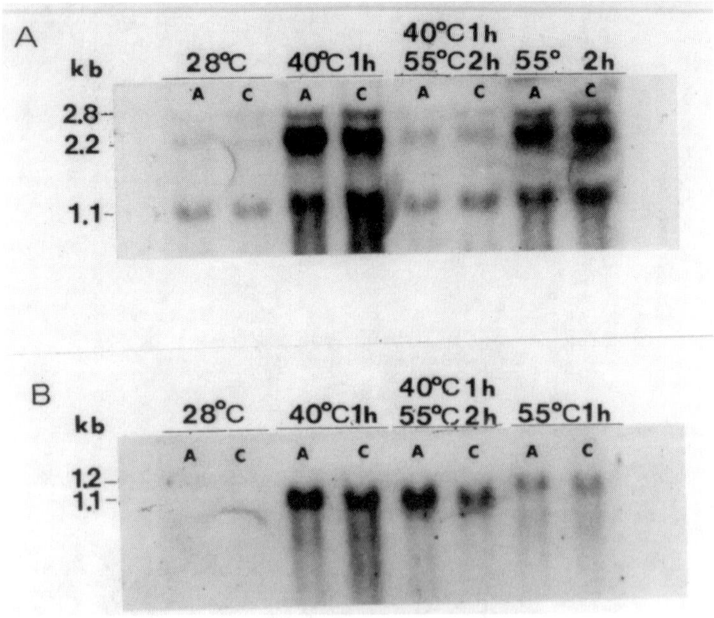

Figure 10.2. Expression of HSP-70 (A) and HSP-26 (B) sequences in perennial ryegrass RNA. Autoradiographs represent the hybridization of total RNA isolated from 4-day-old seedlings of "Accolade" (A) and "Caravelle" (C) after heat treatments as indicated. A control (28°C) that did not experience heat treatment is included. The 20 μg of total RNA per lane was hybridized with pMON-9501 (A) and pZmHSP-26 (B).

mation (Figure 10.2B). Rather, the pZmHSP-26 sequence hybridized to a 1.2 kb transcript in RNA of both cultivars. This message may represent an unprocessed HSP-26 mRNA.

Although HSP-70 gene expression was altered during the heat shock response of perennial ryegrass, there were no detectable differences in the level of HSP-70 mRNA between the two cultivars. Differences in the heat-induced levels of perennial ryegrass HSP-26 sequences correlated with the difference between the two cultivars in response to acclimation temperature and with reported differences in heat tolerance between the two cultivars. Although seedlings exposed to 40°C for 1 hr exhibited no detectable difference in HSP-26 mRNA levels, densitometer measurements revealed the level of the 1.1 kb mRNA in "Accolade" seedlings incubated at 40°C/55°C to be almost twice the level detected in "Caravelle" seedlings exposed to the same temperature regime (Figure 10.2B). Vierling and Nguyen (1992) found similar levels of HSP-26 mRNA in heat-tolerant and heat-sensitive wheat genotypes 1 hr after exposure of seedlings to high temperature. However, when their measurements were taken 2 hr after the initial heat stress, the more thermal-tolerant wheat cultivar had higher levels of HSP-26 mRNA. The increase in level of HSP-26

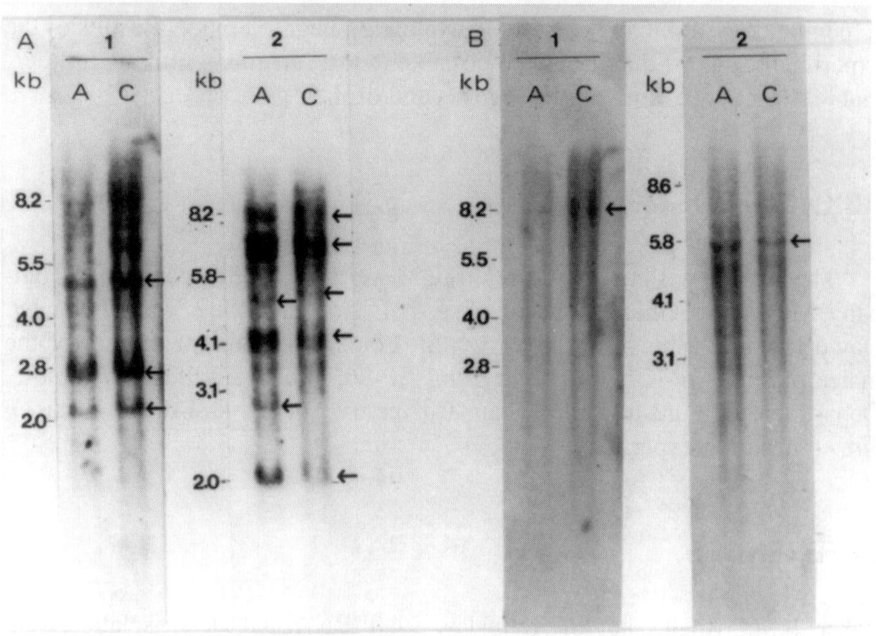

Figure 10.3. Heat shock protein sequence representation in the perennial ryegrass genome. Autoradiographs represent genomic DNA hybridized to HSP-70 (pMON-9501) (A) and HSP-26 (pZmHSP-26) (B) encoding sequences from maize. Ten micrograms of "Accolade" (A) and "Caravelle" (C) DNA restricted with *Eco*R I (filters A1 and B1) and *Hind* III (filters A2 and B2) were loaded per lane.

mRNA detected in "Accolade" relative to "Caravelle" after the 40°C/55°C treatment but not the 40°C treatment may reflect transient differences that occur only after seedlings are incubated under heat stress for a specific period of time.

In vitro translation products differed among RNA extracted from seedlings that experienced treatments of 28°C, 40°C, 40°C/55°C, and 55°C, suggesting that HSP were involved in the acquired thermal tolerance of perennial ryegrass. Results of RNA hybridizations with HSP-26 clones indicated an association between the ability of a heat-tolerant perennial ryegrass cultivar to withstand heat stress and enhanced ability to express at least one HSP gene sequence. It may be possible to exploit this enhanced expression and develop a laboratory screen for improved thermal tolerance. The determination of optimum temperatures and precise timing for HSR induction may be needed in order to develop an accurate screening procedure. Although hybridization of the HSP-70 clone to RNA from the cultivars revealed no differences in level of expression between the two cultivars, RFLP were detected using this clone. Since the RFLP are not likely to be influenced by external conditions of the screening, they too may be useful in evaluating heat tolerance. Results of the experiment suggest that it may be possible to test turfgrass cultivars for thermal tolerance by evaluating HSP genes and/or their gene products.

ACKNOWLEDGMENTS

The authors wish to thank Dr. Virgil Meyer of O.M. Scott and Sons Company, Marysville, Ohio for providing seed of "Accolade" and "Caravelle"; Dr. David Ho of Washington University, St. Louis, Missouri for the gift of the maize plasmids pMON-9501 and pZmHSP-26; and Dr. William Whitmore, Department of Forestry, Ohio State University, for his assistance in the *in vitro* translation experiment.

REFERENCES

Baker, S., C. Rugh, and J.C. Kamalay. 1990. RNA and DNA isolation from recalcitrant plant tissues. *Biotechniques,* 9:268–272.

Baszcynski, C.L., D.B. Walden, and B.G. Atkinson. 1982. Regulation of gene expression in corn (Zea *mays* L.) by heat shock. *Can. J. Biochem.,* 60:569–579.

Beckmann, R.P., L.A. Mizzen, and W. Welch. 1990. Interaction of HSP 70 with newly synthesized proteins: Implications for protein folding and assembly. *Science,* 248:850–854.

Cooper, P. and D. Ho. 1983. Heat shock proteins in maize. *Plant Physiol.,* 71:215–222.

DeRocher, A.E., K.W. Helm, L.M. Lauzon, and E. Vierling. 1991. Expression of a conserved family of cytoplasmic low molecular weight heat shock proteins during heat stress and recovery. *Plant Physiol.,* 96:1038–1047.

DiMascio, J.A., P.M. Sweeney, T.K. Danneberger, and J.C. Kamalay. 1994. Analysis of heat shock response in perennial ryegrass using maize heat shock protein clones. *Crop Sci.,* 34:798–804.

Hendershot, K.L., J. Weng, and H.T. Nguyen. 1992. Induction temperature of heat-shock protein synthesis in wheat. *Crop Sci.,* 32:256–261.

Kimpel, J.A. and J.L. Key. 1985. Heat shock in plants. *Trend. Biochem. Sci.,* 10:353–357.

Mommersteeg International B.V. 1980. "Caravelle" Perennial Ryegrass. U.S. PVP 7700064. Date issued 20 November.

Nieto-Sotelo, J., E. Vierling, and T.D. Ho. 1990. Cloning, sequence analysis, and expression of a cDNA encoding a plastid-localized heat shock protein in maize. *Plant Physiol.,* 93:1321–1328.

O.M. Scott and Sons Company. 1990. "Accolade" Perennial Ryegrass. U.S. PVP 8900297. Date issued 30 April.

Rochester, D.E., J.A. Winer, and D.M. Shah. 1986. The structure and expression of maize genes encoding the major heat shock protein, HSP 70. *EMBO J.,* 5:451–458.

Sullivan, C.Y. and W.M. Ross. 1979. Selecting for drought and heat resistance in grain sorghum. pp. 263–281. In H. Mussell and R.C. Staples (Eds.) *Stress and Physiology in Crop Plants.* Wiley-Interscience, New York.

Vierling, E. 1991. The role of heat shock proteins in plants. *Ann. Rev. Plant Physiol. Mol. Biol.,* 42:579–620.

Vierling, R.A., and H.T. Nguyen. 1992. Heat-shock protein gene expression in diploid wheat genotypes differing in thermal tolerance. *Crop Sci.,* 32:370–377.

Wallner, S.J., M.R. Becwar, and J.D. Butler. 1982. Measurement of turfgrass heat tolerance *in vitro. J. Amer. Soc. Hort. Sci.,* 107:608–613.

White, R.H., P. Stefany, and M. Comeau. 1988. Pre- and post-stress temperature influence on perennial ryegrass *in vitro* heat tolerance. *Hortscience,* 23:1047–1050.

Chapter 11

Development of Transgenic Creeping Bentgrass (*Agrostis palustris* Huds.) for Fungal Disease Resistance

D. Warkentin, B. Chai, A.-C. Liu, R.K. Hajela, H. Zhong, and M.B. Sticklen

INTRODUCTION

Creeping bentgrass (*Agrostis palustris* Huds.) is one of the most widely used turf-type bentgrass species for golf course putting greens, tees, and closely mowed fairways in the USA. It is a cool-season turfgrass in cool and transitional climatic regions and in cooler portions of warm climatic regions, especially in arid zones (Beard, 1982). Its susceptibility to multiple fungal diseases such as brown patch (*Rhizoctonia solani*), *Pythium* blight (*Pythium graminicola*), dollar spot (*Sclerotinia homeocarpa*), and take-all patch (*Gaeumannomyces graminis* var. *avenae*) (Beard, 1973) is a major problem in turfgrass culture. Conventional control of fungal diseases in creeping bentgrass culture often depends on the use of chemically synthesized fungicides, which raises increasing concerns over the environmental impact. Breeding for host resistance to these fungal diseases would be an effective and environmentally sound approach to minimizing the damage to turfgrass culture caused by fungal diseases. It could lead to improvements in the ease and economy of cultural aspects, such as establishment, persistence, durability, and maintenance requirements.

Plant resistance to pathogens involves the accumulation of pathogenesis-related proteins (PR-proteins) that are active in natural defense mechanisms (Collinge and Sluzarenko, 1987). Defense mechanisms involving PR-proteins may sometimes be too weak or too slow to be effective in protecting host plants from pathogens. The progress achieved in the genetic manipulation of plants and the ability to transfer foreign DNA from a variety of sources to plants has facilitated testing the effects of constitutive overexpression of PR-proteins. This may enhance natural defense systems in plants and may significantly extend what could be achieved by traditional breeding methods. Plant transformation using chitinase transgenes for host resistance to fungal diseases has been successfully conducted in tobacco, carrot, and *Brassica napus* (Broglie et al., 1991; Zhu et al., 1994; Stuiver et al., 1996; Grison et al., 1996).

TRANSFORMATION STRATEGIES

Success in regeneration of creeping bentgrass has paved the way for its genetic manipulation. Two types of transformation systems have been tested successfully for developing creeping bentgrass transgenic plants: direct DNA uptake using protoplasts mediated by polyethylene glycol (PEG) or electroporation (Lee et al., 1995); and direct DNA delivery to embryogenic callus via microprojectile bombardment (Zhong et al., 1993; Hartman et al., 1994; Lee et al., 1995; Liu, 1996). The embryogenic callus system is superior over the protoplast system because it is more simplified, less genotype-dependent in regenerability, and prone to less risk of somatic variations, which result from genetic instability as a consequence of the stress of protoplast isolation (Karp, 1994).

To efficiently produce transgenic cells after gene transfer, selectable marker genes are usually introduced for positive selection of transformed cells. A phosphinothricin (PPT) acetyltransferase (PAT) gene has been shown to be an excellent selectable marker in creeping bentgrass transformation (Hartman et al., 1994; Lee et al., 1995; Liu, 1996). Resistance to the herbicide PPT or bialaphos is conferred by PAT, which inactivates PPT by acetylation. PPT or bialaphos selection has also been used successfully in the transformation of potato, tobacco (DeBlock et al., 1987), *Brassica* species (DeBlock et al., 1989), Populus hybrids (DeBlock, 1990), maize (Fromm et al., 1990; Gordon-Kamm et al., 1990) and wheat (Vasil et al., 1992).

Chitinases are PR-proteins that are found in a wide variety of plants. Evidence suggests strongly that chitinases function as antifungal proteins. Accumulation of chitinases (Meins and Ahl, 1989; Rasmussen et al., 1992) and their encoding mRNAs (Metraux and Boller, 1986; Roby and Esquerre-Tugaye, 1987; Meins and Ahl, 1989; Roby et al., 1990) is significantly induced during fungal infection. The induction of chitinase also occurs when plant tissue is treated with fungal cell wall material (Kurosaki et al., 1987; Roby and Esquerre-Tugaye, 1987). Ethylene, a gaseous plant hormone that is normally produced during fungal infection, also induces chitinase activity (Boller et al., 1983; Broglie et al., 1986). In addition, chitinase induction also results from wounding (Parsons et al., 1989). Although chitinases and other PR-proteins have been found to play an active role in natural defense mechanisms, their effectiveness could be greatly compromised or even overcome by pathogens that have evolved mechanisms to evade or inactivate antifungal gene products. As an often used strategy in transformation, foreign genes are chosen as transgenes in order to delay the development of PR-protein tolerance in fungal pathogens. The effective antifungal activity of chitinase genes has been reported in several species of trangenic plants that constitutively produce exogenous chitinases. Transgenic tobacco plants expressing dry bean chitinase were resistant to *Rhizoctonia solani* (Broglie et al., 1991). Transgenic tobacco plants expressing both a rice chitinase and a glucanase gene were also resistant to *Cercospora nicotianae*, the causal agent of frogeye (Zhu et al., 1994). Transgenic carrot plants with both chitinase-I and glucanase-I showed resistance to *Alternaria radicina* and *C. carotae* (Stuiver et al., 1996). Elevated field tolerance to fungal pathogens was first reported in *Brassica napus* transgenic plants that contained a chimeric chitinase gene from bean (Grison et al., 1996).

Genetic transformation of turfgrass has been successful in recent years with the development of efficient regeneration systems and the use of reliable selectable markers and DNA delivery technologies such as biolistic bombardment. However, chitinase genes or other PR-protein genes have not yet been used as transgenes for turfgrass species. Progress in transferring exogenous chitinase genes to turfgrass may greatly benefit turfgrass breeding programs for fungal disease resistance. It would also facilitate our understanding of gene expression and plant-pathogen interactions.

METHODOLOGY

Isolation and Characterization of a Chitinase cDNA Clone

American elm NPS 3-487 (*Ulmus americana*) was selected for its resistance to Dutch elm disease (DED). Callus cultures of NPS 3-487 were initiated from young leaves of greenhouse-grown rooted cuttings. First and second expanded young leaves were surface sterilized in 20% commercial Clorox bleach and cultured in elm callus initiation medium (Sticklen et al., 1986). Cultures were incubated in the dark at 24 ± 2°C for 3–4 weeks until distinct calli (4–5 mm diameter) were established. Vigorously growing calli were subcultured on the same medium every 3 to 4 weeks.

Three avirulent (Q311, Q412, and SSMF) strains and one virulent (CEF16K) strain of *Ophiostoma ulmi* were used for putative induction of elm callus as reported (Hajela and Sticklen, 1993). Fungus-treated callus lines were flash-frozen in liquid nitrogen and stored at 80°C until used.

A unidirectional cDNA library was custom-made by Clontech, Inc., as directed. Briefly, total RNA was isolated from an equal mixture of calli induced as above, using a guanidium-based method (McDonald et al., 1987), and poly A^+ mRNA was affinity purified via poly U Sepharose using standard protocols (Gilmore et al., 1988). Oligo-dT (septadecamer), attached to an Xho I linker, was used as the primer for first strand synthesis. Double-stranded cDNA was linked with EcoRI adapters, and the complete, end-modified cDNA was cloned in EcoRI/Xho I-opened bacteriophage lamba Zap II (Stratagene). The recombinant phage was plated on *E. coli* Sure (Stratagene) cells and amplified one cycle, then frozen in 7% DMSO at 80°C.

A 764-bp wound-inducible *Populus* spp. (poplar) cDNA (Parsons et al., 1989), appearing to encode a chitinase, was used as a heterologous probe at moderate stringency (3X SSPE at 60°C) to screen the cDNA library. This yielded 0.2% primary plaques with varying positive signal intensities. One of the triple plaque-purified phage clones was subcloned into Bluescript II KS (Stratagene) using biological rescue (Short et al., 1988), and sequenced at the Plant Research Laboratory Sequencing Facility at Michigan State University.

Computer-aided sequence analyses of the predicted protein product of pHS2 were conducted on a Macintosh IIsi with either the DNA Strider (shareware version) or Mac Vector software packages. Alignments to known chitinase sequences were performed manually, aided by SeqEd for the Macintosh. Sequence homology searches were performed on the National Center for Biotechnology Information (NCBI) Basic Local Alignment Search Tool (BLAST) network (Altschul et al., 1990).

Genetic Transformation of Creeping Bentgrass

Plant Material and Culture

Embryogenic callus was induced from mature seeds (caryopses) of creeping bentgrass (*Agrostis palustris* Huds.) cultivar Penncross following the methods established by Zhong et al. (1993). Seeds were first surface sterilized with 50% commercial Clorox bleach solution containing 1% Tween 20 with a vacuum applied and then soaked in 70% ethanol for 5 minutes. The sterilized seeds were finally rinsed with sterilized distilled water three times before being transfer to the semi-solid medium containing MS basal salts supplemented with 500 mg/L enzymatic casein hydrolysate, 3% sucrose, 30 μM Dicamba (3,6-dicloro-*o*-anisic acid), 2.25 μM BA (6-benzyladenine), and 7 g/L phytagar. Cultures were incubated in the dark at 25°C and subcultured every two weeks. Light yellow, friable calli were selected and placed on a 2-cm^2 area in a single layer in the center of Petri dishes containing callus induction medium for microprojectile bombardment.

Microprojectile Bombardment

A Biolistic PDS-1000/He system (DuPont/Bio-Rad) was used to deliver tungsten particles coated with plasmid DNA. Physical parameters were optimized to increase the numbers of transiently GUS-expressing cells. The following conditions were found to be superior and used as a standard bombardment protocol: rupture disk pressure, 1,550 psi; gap distance from rupture disk to macrocarrier, 6 mm; macrocarrier travel distance, 16 mm; microcarrier travel distance, 6 cm. Prior to bombardment, plasmid DNA was precipitated onto tungsten particles (0.9–1.2 μm in diameter) following the protocol described by Zhong et al. (1993), using a precipitation mixture that included 1.5 mg tungsten, 30 μg plasmid DNA, 1.1 *M* $CaCl_2$ and 8.7 m*M* spermidine (free base). The plasmid construct pJS101 carrying the selectable marker coding sequence (*bar* gene) driven by the CMV 35S promoter was provided by Dr. Ray Wu of Cornell University, while the pHS2 gene (American elm chitinase gene) driven by the CMV 35S promoter is harbored in the plasmid vector pKYLX-71 (Schardl et al., 1987). To optimize the conditions for co-transformation, the pHS2-harboring plasmid DNA and the *bar* gene-harboring plasmid DNA were coprecipitated in different ratios. Embryogenic callus was bombarded three times and transferred to fresh medium immediately after bombardment to minimize contamination.

Selection for Transformants

Selection was conducted in three stages: callus, seedlings, and greenhouse plants with an increasing concentration of the selective agent. Two weeks after bombardment, calli were subcultured onto selection medium containing 3 mg/L bialophos. Additional selections at the callus stage were conducted at a 4-week subculture interval under 5 mg/L and 10 mg/L bialophos in consecutive order. The surviving calli on 10 mg/L bialophos were transferred to MS medium containing 10 mg/L bialophos and incubated under lights for regeneration. Thousands of these putative transformed plantlets were regenerated via somatic embryogenesis and were transferred to the greenhouse one month after culturing in regeneration medium. Selection in the greenhouse was done by spraying the putatively transformed plants with 1% Ignite®* solution, containing 180 mg/L PPT. Two sprays were applied 2 weeks and 8 weeks after transfer to the greenhouse. The effect of sprays was evaluated 7 days after herbicide application.

---

* Registered Trademark of Hoechst-Roussel Agri-Vet Company, Leland, MS.

RESULTS AND DISCUSSION

Isolation and Characterization of a Chitinase cDNA Clone

A full-length, 1225-bp cDNA clone containing a 951-nucleotide open reading frame (ORF) was isolated. Using the deduced 317-amino acid sequence from this ORF, a homology search using BLAST showed 50 of the 53 best matches to be with chitinases; the remaining three were lectins or agglutinins, which share a chitin-binding domain with the pHS2 translation product. This product revealed strong homology to bean, poplar, and tobacco class I chitinases, strongly suggesting that this elm cDNA encodes a chitinase. For this reason, we designated the predicted protein product of this clone ECH2 (elm chitinase 2). The deduced amino acid seaquake of ECH2 has 71.2% sequence identity to the *Phaseolus vulgaris* chitinase clone pCH18 (Broglie et al., 1986), 70.5% to the tobacco CHN50 chitinase (Shinshi et al., 1990), and 70% to the poplar WIN6 (Parsons et al., 1989) translation product.

The 3′ untranslated region of pHS2 reveals three putative polyadenylation signals as described by Joshi (1987). A classic AATAAA signal is not present; however, an AATAAG and an AACAAA are present. Currently, we believe the AATAAG to be the more probable signal, as this motif alone is present in another chitinase-like clone from this elm (unpublished results). Immediately 5′ to the AATAAG is a YAYTG-like sequence (CAATG). Finally, ten nucleotides downstream, a TGTGTGCACT is present with high identity to a third polyadenylation motif (Joshi, 1987).

ECH2 has an overall charge of 10 and a deduced pI of 8.49. The first 21 residues of ECH2 comprise a signal peptide; this peptide is hydrophobic and has a positively charged residue proximal to the initial methionine (von Heijne, 1987). Residues 314–321 appear to include a second routing peptide known as a C-terminal extension. It has been established that similar C-terminal extensions are sufficient and necessary to direct a protein into the vacuole (Chrispeels and Raikhel, 1993). Residues 22–313 compose the main, catalytic domain.

Transformation of Creeping Bentgrass

We have regenerated thousands of putative transformed plantlets from seven out of twelve bialophos-resistant callus colonies. All of the calli and plantlets from these colonies are consistent in resistance to increasing concentrations of bialophos. Molecular confirmation of the presence and expression of pHS2 gene in the transgenic plants is in progress. Our goal is to generate transformants that have integrated the pHS2 gene in their genomes and to evaluate the efficiency of cotransformation as a function of the coprecipitation ratios of the two plasmids that carry the target gene (pHS2) and the selectable marker gene (*bar*). Polymerase chain reaction (PCR) will be used to screen bialophos-resistant lines for those that have pHS2 insertions in their genomes.

Expression of pHS2 will be studied by examining the gene products, mRNA, and ECH2 protein. Bioassays for fungal disease resistance will be conducted *in vitro* to study the effect of the isolated chitinase on fungal pathogens. Studies *in vivo* will be conducted to assay symptom development in transgenic plants inoculated with fungal pathogens.

REFERENCES

Altschul, S.F., W. Gish, W. Miller, E.W. Myers, and D.J. Lipman. 1990. Basic local alignment search tool. *J. Mol. Biol.*, 215:403–410.

Beard, J.B. 1973. *Turfgrass: Science and Culture*. Prentice-Hall, Inc., Englewood Cliffs, NJ. pp. 71–78.

Beard, J.B. 1982. *Turf Management for Golf Courses*. Macmillan, NY, pp. 119–124.

Boller, T., A. Gheri, F. Mauch, and U. Vogeli. 1983. Chitinase in bean leaves: induction by ethylene, purification, properties, and possible function. *Planta.*, 157:22–31.

Broglie, K.E., J.J. Gaynor, and R.M. Broglie. 1986. Ethylene-regulated gene expression: molecular cloning an endochitinase from *Phaseolus vulgaris. Proc. Nat. Acad. Sci. USA.,* 83:6820–6824.

Broglie, K.E., I. Chet, M. Holliday, R. Cressman, P. Biddle, S. Knowlton, C.J. Mauvais, and R. Broglie. 1991. Transgenic plants with enhanced resistance to the fungal pathogen *Rhizoctonia solani. Science,* 254:1194–1197.

Chrispeels, M.J. and N.Y. Raikhel. 1993. Short peptide domains target proteins to plant vacuoles. *Cell,* 68:613–616.

Collinge, D.B. and A.J. Sluzarenko. 1987. Plant gene expression response to pathogens. *Plant Mol. Biol.*, 9:389–410.

DeBlock, M. 1990. Factors influencing the tissue culture and the *Agrobacterium tumefaciens*-mediated transformation of hybrid aspen and poplar clones. *Plant Physiol.*, 93:1110–1116.

DeBlock, M., J. Botterman, M. Vandewiele, J. Dockx, C. Thoen, V. Gossele, N.R. Mowa, C. Thompson, M. van Montagu, and J. Leemans. 1987. Engineering herbicide resistance in plants by expression of a detoxifying enzyme. *EMBO J.*, 6:513–2518.

DeBlock, M., D. Brouwer, and P. Tenning. 1989. Transformation of *Brassica napus* and *Brassica oleracea* using *Agrobacterium tumefaciens* and the expression of the *bar* and *neo* genes in the transgenic plants. *Plant Physiol.*, 91:694–701.

Donn, G., M. Nilges, and S. Morocz. 1996. Stable transformation of maize with a chimeric, modified phosphinothricin-acetyltransferase gene from *Streptomyces virido-chromogenes. Abstracts 7th Intl. Congress on Plant Tissue and Cell Culture*. Abstract No. A2-38, p. 63.

Fromm, M.E., F. Morris, C. Armstrong, R. Williams, J. Thomas, and T.M. Klein. 1990. Inheritance and expression of chimeric genes in the progeny of transgenic maize plants. *Bio/Technology*, 8:833–839.

Gilmore, S.J., R.K. Hajela, and M.F. Thomashow. 1988. Cold acclimation in *Arabidopsis thaliana. Plant Physiol.*, 87.745–750.

Gordon-Kamm, W.J., T.M. Spencer, M.L. Mangano, R.J. Daines, W.G. Start, J.V. O'Brien, S.A. Chambers, W.R. Adams, Jr., N.G. Willetts, T.R. Rice, C.J. Mackey, R.W. Krueger, A.P. Karusch, and P.G. Lemaux. 1990. Transformation of maize cells and regeneration of fertile transgenic plants. *The Plant Cell,* 2:603–618.

Grison, R., B. Grezes-Besset, M.Schneider, N. Lucante, L. Olsen, J.J. Leguay, and A. Toppan. 1996. Field tolerance to fungal pathogens of *Brassica napus* constitutively expressing a chimeric chitinase gene. *Nature Biotechnology*, 14:643–646.

Hajela, R.K. and M.B. Sticklen. 1993. Cloning of pathogenesis-related genes from *Ulmus americana*, in, *Dutch elm disease research: cellular and molecular approaches*. Sticklen, M.B. and J.L. Sherald, Eds. Springer-Verlag, NY. pp. 193–207.

Hartman, C.L., L. Lee, P.R. Day, and N.E. Tumer. 1994. Herbicide resistant turfgrass (*Agrostis palustris* Huds.) by biolistic transformation. *Bio/Technology,* 12:919–923.

Joshi, C.P. Putative polyadenylation signals in nuclear genes of higher plants: a compilation and analysis. *Nucl. Acids Res.*, 15:9627–9640, 1987.

Karp, A. 1994. Origins, causes and uses of variation in plant tissue culture. In *Plant Cell and Tissue Culture*, Vasil, I.K. and T.A. Thorpe, Eds., Kluwer Academic Publishers. pp. 139–151.

Kurosaki, F., N. Tashihiro, and A. Nishi. 1987. Secretion of chitinase from cultured carrot cells treated with fungal mycelial walls. *Physiol. Mol. Plant Pathol.*, 31:211–216.

Lee, L., C. Hartman, C. Laramore, N. Tumer, and P. Day. 1995. Herbicide-resistant creeping bentgrass. *USGA Green Section record.* 33(3):16–18.

Liu, C.A. 1996. Development of herbicide-resistant turfgrass via biolistic gene genetic engineering. Dissertation. Michigan State University. 187 pp.

McDonald, R.J., G.H. Swift, A.E. Przybyla, and J.M. Chirgwin. 1987. Isolation of RNA using guanidinium salts. *Methods Enzymol.*, 152:217–227.

Meins, F. and P. Ahl. 1989. Induction of chitinase and 1,3-glucanase in bean leaves. *Plant Cell,* 1:447–457.

Metraux, J.P. and T. Boller. 1986. Local and systemic induction of chitinase in cucumber plants in response to viral, bacterial, and fungal infections. *Physiol. Mol. Plant Pathol.,* 28:161–169.

Parsons, T.J., H.D. Bradshaw, and M.P. Gordon. 1989. Systemic accumulation of specific mRNAs in response to wounding in poplar trees. *Proc. Nat. Acad. Sci. USA*, 86:7895–7899.

Rasmussen, U., H. Geise, and J.D. Mikkelsen. 1992. Induction and characterization of chitinase in *Brassica napus* L. spp. *oleifera* infected with *Phoma lingam. Planta,* 187:328–334.

Roby, D. and M.T. Esquerre-Tugaye. 1987. Chitin oligosaccharides as elicitors of chitinase activity in melon plants. *Biochem. Biophys. Res. Commun.*, 143:885–892.

Roby, D., K. Broglie, R. Cressman, P. Biddle, and I. Chet. 1990. Activation of a bean chitinase promoter in transgenic tobacco plants by phytopathogenic fungi. *Plant Cell.*, 2:999–1007.

Schardl, C.L., D.B. Alfred, B. Gary, A.A. Mitchell, F.H. David, and G.H. Arthur. 1987. Design and construction of a versatile system for the expression of foreign genes in plants. *Gene.,* 61:1–11.

Shinshi, H., J.M. Neuhaus, J.M. Ryals, and F. Mains. 1990. Structure of a tobacco endochitinase gene: evidence that different chitinase genes can arise by transposition of sequences encoding a cysteine-rich domain. *Plant Mol. Biol.*, 14:357–368.

Short, J.M., J.M. Fernandez, J.A. Sorge, and W.D. Huse. 1988. ZAP, a bacteriophage expression vector with *in vivo* properties. *Nucl. Acids Res.*, 16:7583–7600.

Sticklen, M.B., S.C. Domit, and R.D. Lineberger. 1986. Shoot regeneration from protoplasts of *Ulmus* Pioneer. *Plant Sci.*, 47:29–34.

Stuiver, M.H., J.B. Bade, H. Tigelaar, L. Molendijk, E. Troost-van Deventer, M.B. Sela-Buurlage, J. Storms, L. Plooster, F. Sijbolts, J. Custers, M. Apothekerde Groot, and L.S. Melchers. 1996. Broad spectrum fungal resistance in transgenic carrot plants. *Abstract of 1996 Meeting of the Society for In Vitro Biology Biotechnology*. SP-1014, p. 14.

Vasil, V., A.M. Castillo, M.E. Fromm, and I.K. Vasil. 1992. Herbicide resistant fertile transgenic wheat plants obtained by microprojectile bombardment of regenerable callus. *Bio/Technology*, 10:667–674.

von Heijne, G. 1987. Patterns of amino acids near signal-sequence cleavage sites. *Eur. J. Biochem.*, 133:17–21.

Zhong, H., C. Srinivasan, and M.B. Sticklen. 1991. Plant regeneration via somatic embryogenesis in creeping bentgrass (*Agrostis palustris* Huds). *Plant Cell Rep.*, 10:453–456.

Zhong, H., M.G. Bolyard, C. Srinivasan, and M.B. Sticklen. 1993. Transgenic plants of turfgrass (*Agrostis palustris* Huds) from microprojectile bombardment of embryogenic callus. *Plant Cell Rep.*, 13:1–6.

Zhu, Q., E.A. Maher, S. Masoud, R.A. Dixon, and C.J. Lamb. 1994. Enhanced protection against fungal attack by constitutive-co-expression of chitinase and glucanase genes in transgenic tobacco. *Bio/Technology,* 12:807–812.

Part 4

In Vitro Culture and Genetic Engineering of Turfgrass

Chapter 12

Utilizing *In Vitro* Culture for the Direct Improvement of Turfgrass Cultivars

I. Yamamoto and M.C. Engelke

ABSTRACT

Developing pest-resistant cultivars is a highly desirable alternative, both environmentally and economically, to using present-day pesticide programs. Pest resistance can be efficiently introduced into turfgrass cultivars by combining cell biology and molecular genetic techniques with conventional breeding methods. Most of the recent progress in turfgrass tissue culture systems used embryo (seed)-derived callus. Since seeds are genetically heterogeneous in cross-pollinated cultivars (e.g., most of the creeping bentgrasses), plants regenerated from a seed-derived callus are genetically different from each other and from the parental lines, in addition to any possible somaclonal variations. Therefore, extensive selection processes, based on agronomic characteristics, are still required before regenerated or transformed plants are released as new cultivars. Development of a cell-culture system that possesses the genetic information of superior cultivars or lines could be the most straightforward, efficient, and desirable approach. To improve existing cultivars, we are trying to develop an *in vitro* culture system using nodal segments for six parental clones of 'Crenshaw' creeping bentgrass (*Agrostis palustris* Huds.) and five zoysiagrass lines (*Z. japonica* and *Z. matrella*). The use of somatic tissue from parental clones of cross-pollinated cultivars or vegetatively propagated cultivars could allow us to introduce a new gene(s) directly into preexisting superior cultivars. The development of an *in vitro* culture system using nodal segments or crowns could thus be a practical breeding tool for cross-pollinated as well as vegetatively propagated turfgrass cultivars.

INTRODUCTION

The incidence of disease and/or insect infestation has often complicated turfgrass management. The primary cultural practice for pest control centers on the timely, and voluminous, use of pesticides. Over the past few years, however, considerable emphasis has been placed on alternative forms of pest management, such as the breeding of pest-resistant cultivars. More recently, successful biotechnology applications in the turfgrass area suggest that we can efficiently introduce pest resistance into turfgrass cultivars by combining cell biology and molecular genetic techniques with conventional breeding methods. However, most of these applications have been limited to the use of an embryo (seed)-derived callus (Al-Khayri et al., 1989; Asano, 1989; Asano and Ugaki, 1994; Blanche et al., 1986; Ha et al., 1992; Hartman et al., 1994; Kuo et al., 1994; Krans et al., 1982; Terakawa et al., 1992; Zhong et al., 1991; Zhong et al., 1993). In cross-pollinated cultivars (e.g., most of the creeping bentgrasses), seeds are genetically heterogeneous; therefore, the plants regenerated from a seed-derived callus are genetically different from each other and from the parental lines, in addition to any possible somaclonal variations. Thus, extensive selection processes based on the agronomic characteristics are still required before an *in vitro* selection or transgenic plants are released as new cultivars.

Our research efforts have focused on the development of a tissue-culture system that possesses the genetic information of superior cultivars or lines. We are trying to induce embryogenic callus derived from nodal segments of the six parental clones of 'Crenshaw' creeping bentgrass (*Agrostis palustris* Huds.) and five zoysiagrass lines *(Z. japonica* and *Z. matrella*). The approach is a pragmatic breeding tool for improving existing cross-pollinated and vegetatively propagated turfgrass cultivars.

TURFGRASS BREEDING AND APPLICATIONS OF MOLECULAR GENETIC TECHNIQUES

Cross-Pollinated Cultivars

Most of the recently released cool-season turfgrasses, except Kentucky bluegrass (*Poa pratensis* L.), are synthetic cultivars (Burton, 1992). In synthetic cultivars, turfgrass is evaluated by the performance of millions of its siblings, not an individual plant. When one of the million seeds is used as an explant for an *in vitro* culture system, it is unknown whether the explant possesses agronomically or physiologically desirable genes. For example, when individual seedlings are grown separately, a large degree of variation in morphology or physiology can be found within a cultivar. Therefore, when embryos are used as explants there is no guarantee that the explants possess desirable characteristics. Consequently, after the successive achievement of transgenic plants or cell line selections, a completely new set of conventional breeding efforts is required before biotechnologically produced plants can be used commercially. Alternatively, if we manipulate the parental plants of existing cultivars or already selected elite lines, the progeny produced from the manipulated parents can be released as new improved cultivars without any extensive selection processes.

Crenshaw creeping bentgrass is a six-clone synthetic cultivar released in 1993 by the Texas Agricultural Experimental Station (TAES) at Dallas. The parental clones (TAES2737, 2739, 2740, 2741, 2743, and 2895) originated from Arizona and Texas, and three of them (TAES2739, 2740, and 2743) were also used as parental clones for a five-clone synthetic cultivar, "SR1020," released by University of Arizona in 1988. Crenshaw has been identified as one of the premier bentgrasses for the golf industry. We can generally attribute this accolade to its superior performance under heat and drought conditions and its tolerance to *Rhizoctonia* blight and *Pythium* blight diseases. Its Achilles heel, however, is its susceptibility to *Sclerotinia* dollar spot (*S. homeocarpa*) (Colbaugh et al., 1993), which has restricted foliar blighting activity and raises cosmetic concerns on bentgrass putting greens (Smiley et al., 1992). Insertion of the *S. homeocarpa*–resistant gene into Crenshaw without changing any other characteristics, for example, is a resourceful way to capitalize on advanced genetic techniques for turfgrass breeding.

We have conducted several screening tests to identify the relative susceptibility (tolerance) among Crenshaw parental clones to *S. homeocarpa*. The experiments revealed variable responses to the organism (unpublished data), suggesting that the incorporation of a resistant gene(s) into the susceptible parental lines might ultimately produce the *S. homeocarpa*–resistant (tolerant) Crenshaw. The recent successes in the development of transgenic turfgrasses (Ha et al., 1992; Hartman et al., 1994; Zhong et al., 1991) suggest the potential for such an approach. Alternative breeding schemes (conventional methods) can be employed; however, the genetic engineering approach may be fully justified as consistent with the single trait improvement of Crenshaw.

Vegetatively Propagated Cultivars

Most of the preexisting warm-season turfgrasses are released as vegetative cultivars (Burton, 1992). A vegetative cultivar already possesses desirable and specific characteristics; therefore, the direct manipulation of the elite clones (cultivars) by molecular genetic techniques should be an easy and efficient way to improve existing cultivars. Plant regeneration via callus (Al-Khayri et al., 1989) and protoplast (Asano, 1989) cultures was reported in zoysiagrasses using mature caryopses (embryos) as explants. In St. Augustinegrass [*Stenotaphrum secundatum* (Walt.) Kuntze] immature embryos were used to induce callus and regenerate plants (Kuo and Smith, 1993). On the other hand, bermudagrass [*Cynodon dactylon* (L.) Pers.] was regenerated from immature inflorescence-derived callus (Ahn et al., 1985). This is one of the limited studies using non-embryo tissues as explants in turfgrass species. Inflorescences are potential explants for other warm-season species studies. However, inflorescence may not be available through all seasons, and some cultivars may not produce a sufficient number of flowers. Thus, the development of a tissue culture system using nodal segments (meristematic tissues) could be instrumental for efforts to apply advanced molecular genetic techniques to breeding of vegetatively propagated turfgrasses.

Five elite zoysiagrass lines, 'Diamond (*Z. matrella*),' 'Cavalier (*Z. matrella*),' and 'Crowne (*Z. japonica*),' recently released by TAES-Dallas, and TAES experimental lines, 'DALZ8501 (*Z. matrella*)' and 'DALZ8516 (*Z. japonica*),' were selected as experimental genotypes. These lines were developed, based on various morphological and physiological characteristics, by conventional breeding methods. Although the lines possess several desirable agronomic advantages as turf, some disadvantages also exist. For example, although Cavalier is resistant to important pests of zoysiagrass in the southern United States, such as the tropical sod webworm (TSW) (*Herpetogramma phaeopteralis*) and the fall armyworm (*Spodoptera frugiperda*), it is susceptible to zoysia mites (*Eirophyes zoysiae*) (Reinert, 1992). Diamond, which possesses better agronomic characteristics than DALZ8501, is susceptible to TSW and zoysia mites. On the other hand, DALZ8501, a line that morphologically resembles Diamond, is highly resistant to TSW and moderately resistant to zoysia mites.

The introduction of specific pest resistances into these lines can enhance turf performance and indirectly reduce chemical dependence. We can transfer the insect-resistant gene(s) from resistant lines to susceptible lines by conventional recurrent selection techniques; however, the recent advanced molecular genetic techniques may accomplish the transfer within a much shorter time period.

THE DEVELOPMENT OF TISSUE CULTURE SYSTEMS FOR THE IMPROVEMENT OF TURFGRASS CULTIVARS

Materials and Methods

Stolons of parental clones of Crenshaw creeping bentgrass and stolons or rhizomes of zoysiagrass elite lines were harvested from plant materials grown under greenhouse conditions. For zoysiagrasses, we selectively harvested stolons from the sections that did not touch the soil surface. The soil medium was a 2:3 volume mixture of sterilized sand and Metro-Mix 200.®* We prewashed stolons or rhizomes with running water, agitated in a 0.1% (v/v) detergent solution, and rinsed under running distilled water. From the prewashed stolons or rhizomes, we harvested 2- to 3-mm nodal segments for bentgrasses and 5- to 7-mm nodal segments for zoysiagrasses. The segments were surface sterilized in a 1% (v/v) mild liquid detergent solution for 20 min, rinsed under running distilled water for 10 min, and soaked in a 25% (v/v) commercial bleach solution (≈1.3% NaOCl) plus 0.1% (v/v) surfactant for 10 min. (Smith et al., 1993). The explants were then rinsed four to six times with sterilized distilled water.

The culture medium for callus induction was Murashige and Skoog (MS) basal media with 1.0 mg/L MS vitamins, 30 g/L sucrose, 1.0 g/L $MgCl_2 \cdot 6H_2O$, and 1.0 to 3.0 mg/L of 2,4-dichlorophenoxyacetic acid (2,4-D). For a plant regeneration medium, a half-strength (1/2) MS media without 2,4-D was used. In some experiments, 0.5 to 3.0 mg/L 6-benzylaminopurine (BA) was added into the regeneration medium. The pH was adjusted to 5.8 with 2 or 1 *M* NaOH. The medium was solidified with 7.7 g/L of Phytagel®** and autoclaved at 125°C for 15 min. All cultures were maintained at 25 ± 2°C under fluorescent light for 8 hr/day and subcultured every 4 to 6 weeks.

The surface sterilized nodal segments were also cultured directly into the regeneration medium. After contamination-free nodal segments were selected and shoots and roots emerged from them, we transplanted them to 5-cm (diameter) by 8-cm jars for further growth. The growth media were 1/2 MS medium or 1/2 MS medium plus 0.5 mg/L BA. From the propagated plants, nodal segments were cut off and then used as secondary explants.

Preliminary Results and Discussion

According to Ahn et al. (1985), bermudagrass was regenerated from a callus derived from immature inflorescence tissue. Thus, immature inflorescence could be a potential candidate for explant tissue parts, since most of the zoysiagrasses form inflorescence under natural conditions in the southern United States. Nevertheless, in our study, we selected nodes as explants since they are readily available throughout the year. However, one disadvantage of nodal segment use is the difficulty of obtaining contamination-free explants. We selected the sterilization method previously mentioned over the ethanol-NaOCl methods commonly used, since it provided better results. Despite this,

---

* Metro-Mix 200 is a product of Scott-Sierra Horticultural Products Co., Marysville, Ohio. The use of trademarks, proprietary product, or vendor does not constitute a guarantee or warranty for the products and does not imply its approval to the exclusion of other products or vendors that may be suitable.

** Gellan gum, agar substitute gelling agent by Sigma Chemical Co., St. Louis, Missouri.

the average percentage of contamination-free explants was still only 25.9% for bentgrasses and around 17.7% for zoysiagrasses when we took nodes from stolons. Moreover, this percentage tended to decrease when nodes were harvested from rhizomes. To overcome this problem, we attempted to produce contamination-free nodes by culturing sterilized explants *in vitro*. The surface sterilized explants were cultured directly into the regeneration medium, and contamination-free explants successfully obtained were transplanted to 1/2 MS medium jar after the emergence of shoots and roots. The nodal segments harvested from the grown-up plants (secondary explants) were 100% contamination-free and no additional sterilization processes were required. The use of a solution culture before the harvest could be another way to resolve the contamination problem.

We observed the first and best callus induction 9 to 13 days after incubation for bentgrasses when cultured in MS medium with 1.0 mg/L 2,4-D. The average callus induction rates were 80.8% to 91.1%, depending on genotypes. Other media evaluated were the MS medium with 1.0 mg/L 3,6-dichloro-2-methoxybenzoic acid (dicamba) and Nitsch and Nitsch (N-N) basal salt medium with 1.0 mg/L 2,4-D. After successive three to four subcultures in the MS medium, at 4- to 6-week intervals, calli were transferred into the 1/2 MS plant regeneration medium. Plants were regenerated from opaque, white to yellowish compact calli. The regenerated plants were similar to each other and to the explant clones. To examine somaclonal variations, genetic examinations will eventually be employed.

The secondary explants also induced calli. Thus, the contamination-free secondary explants facilitated our efforts to determine adequate culture systems or environments for callus induction and plant regeneration. As in the case of the original nodal segments, plants were regenerated after the calli were transferred to the 1/2 MS medium.

The nodal segments of *Z. japonica* induced callus 7 to 8 days after incubation in the MS medium with 1.0 mg/L 2,4-D, while it took 14 to 21 days for *Z. matrella*. The frequencies of callus induction ranged from 61.1% to 92.9%, but the growth of callus was somewhat slow. This suggests that the modification of culture and/or environmental conditions is necessary. To date, several different types of calli were developed and were under selection for their plant regeneration ability.

CONCLUSION

We have presented some preliminary results from an ongoing project. Nevertheless, these early results strongly suggest that nodal segments can be used as explants for inducing calli that regenerate plants. For bentgrasses, as in the case of seed-derived callus, regeneration of plants from embryogenic callus was observed. The use of nodal segments as explants, and the development of its accompanying tissue culture system, will facilitate the incorporation of biotechnology into turfgrass breeding.

REFERENCES

1. Al-Khayri, J., F.H. Huang, L.F. Thompson, and J.W. King. 1989. Plant regeneration of zoysiagrass from embryo-derived callus. *Crop Sci.,* 29:1324–1325.
2. Ahn, B.J., F.H. Huang, and J.W. King. 1985. Plant regeneration through somatic embryogenesis in common bermudagrass tissue culture. *Crop Sci.,* 25:1107–1109.
3. Asano, Y. 1989. Somatic embryogenesis and protoplast culture in Japanese lawngrass (*Zoysia japonica*). *Plant Cell Rep.,* 8:141–143.
4. Asano, Y. and M. Ugaki. 1994. Transgenic plants of *Agrostis alba* obtained by electroporation-mediated direct gene transfer into protoplasts. *Plant Cell Rep.,* 13:243–246.
5. Blanche, F.C., J.V. Krans, and G.E. Coats. 1986. Improvement in callus growth and plantlet formation in creeping bentgrass. *Crop Sci.,* 26:1245–1248.
6. Burton, G.W. 1992. Breeding improved turfgrass. pp. 759–776. In D.V. Waddington, R.N. Carrow, and R.C. Shearman (Ed.) *Turfgrass Agron. Monogr.* 32. ASA, CSSA, and SSSA, Madison, WI.
7. Colbaugh, P.F., S.P. Metz, and M.C. Engelke. 1993. *Sclerotinia* dollar spot incidence on bentgrass. *Texas Turf Rep.* CPR 5104-5146:88.
8. Ha, S.-B., F.-S. Wu, and T. K. Thorne. 1992. Transgenic turf-type tall fescue (*Festuca arundinacea* Schreb.) plant regenerated from protoplast. *Plant Cell Rep.,* 11:601–604.
9. Hartman, C.L., L. Lee, P.R. Day, and N.E. Tumer. 1994. Herbicide resistant turfgrass (*Agrostis palustris* Huds.) by biolistic transformation. *Bio. Tech.,* 12:919–923.
10. Kuo, Y.-J. and M.A.L. Smith. 1993. Plant regeneration from St. Augustinegrass immature embryo-derived callus. *Crop Sci.,* 33:1394–1396.
11. Kuo, Y.-J., M.A.L. Smith, and L.A. Spomer. 1994. Merging callus level and whole plant microculture to select salt-tolerant 'Seaside' creeping bentgrass. *J. Plant Nut.,* 17:549–560.
12. Krans, J.V., V.T. Henning, and K.C. Torres. 1982. Callus induction, maintenance and plantlet regeneration in creeping bentgrass. *Crop Sci.,* 22:1193–1197.
13. Reinert, J.A. and M.C. Engelke. 1992. Resistance in zoysiagrass (*Zoysia* spp.) to the tropical sod webworm (*Herpetogramma phaeopteralis*). *Texas Turf Rep.* PR 4996:54–55.
14. Smiley, R.W., P.H. Dernoeden, and B.B. Clarke. 1992. Compendium of turfgrass diseases, 2nd ed. APS Press, St. Paul, MN.
15. Smith, M.A., J.E. Meyer, S.L. Knight, and G.S. Chen. 1993. Gauging turfgrass salinity responses in whole-plant microculture and solution culture. *Crop Sci.,* 33:566–572.

16. Terakawa, T., T. Sato, and M. Koike. 1992. Plant regeneration from protoplasts isolated from embryogenic suspension cultures of creeping bentgrass (*Agrostis palustris* Huds.). *Plant Cell Rep.,* 11:457–461.
17. Zhong, H., C. Srinivasan, and M.B. Sticklen. 1991. Plant regeneration via somatic embryogenesis in creeping bentgrass (*Agrostis palustris* Huds.). *Plant Cell Rep.,* 10:453–456.
18. Zhong, H., M.G. Bolyard, C. Srinivasan, and M.B. Sticklen. 1993. Transgenic plants of turfgrass (*Agrostis palustris* Huds.) from microprojectile bombardment of embryogenic callus. *Plant Cell Rep.,* 13:1–6.

Chapter 13

Embryo Production in Orchardgrass

H. Brittain-Loucas, B. Tar'an, S.R. Bowley, B.D. McKersie, and K.J. Kasha

INTRODUCTION

Orchardgrass (*Dactylis glomerata* L.) is an autotetraploid perennial grass used as a forage crop. With the development of artificial seed technology, it may be feasible to use this system to propagate parents for commercial production and release of hybrid varieties. Furthermore, if parents of a variety can be manipulated *in vitro*, this would facilitate the creation of varieties with value-added traits introduced by genetic transformation. Prerequisites necessary to apply these technologies include: (1) germplasm with commercial utility that can form *in vitro* somatic embryos; and (2) tissue culture protocols that will produce high numbers of somatic embryos synchronized for maturity.

Cultivated varieties, like most perennial grasses, are the progeny of a selected population of several parents. These populations, termed synthetic varieties, are increased in isolation for three or more generations of seed increase in order to obtain sufficient seed of commerce. Inbreeding that occurs during the seed increase steps decreases the performance such that the yield of the seed of commerce (certified generation) is often significantly below that of the first, or breeder, generation. F_1 and Double-cross hybrids have been developed with yield, quality, and persistence attributes that are superior to synthetic varieties. However, traditional methods used for propagation of parents for commercial seed production are not cost-effective, and this has prevented their introduction to the market. *In vitro* cultures are capable of producing large numbers of somatic embryos, which are genetic copies of the original plant. With the development of artificial seed technology, it may be feasible to use this system to propagate parents for commercial production and release of hybrid varieties (McKersie and Bowley, 1993).

If parents of a variety can be manipulated *in vitro*, this would facilitate the creation of varieties with value-added traits introduced by genetic transformation. By transforming plants of commercial utility, one would reduce or possibly eliminate the amount of conventional breeding required to produce a commercial, genetically engineered variety. Genes to enhance existing attributes or to confer new traits not exhibited in the species can be introduced during *in vitro* culture of the parental plants. The desired transgenic form of the parental plants can be identified and the utility of the introduced gene in a commercial variety assessed by evaluating the progeny of these plants.

IN VITRO CULTURE

Commercial varieties of orchardgrass, like many other species, produce little or no somatic embryos using current techniques. One individual genotype, Regen-O, was selected from within the variety Potomac for its high capacity to form somatic embryos from mesophyll cells (Conger and Hanning, 1991). A flow diagram of the *in vitro* procedures is presented in Figure 13.1. Explants from leaf bases are cultured in the dark at 25°C on solid SH medium with 30 μ*M* dicamba (SH-30 medium). Explants are obtained by excising the innermost leaf of vegetative tillers, which is then split longitudinally down the midrib. Leaves are surface sterilized in 2% alcotabs (Alconox Inc.) as a surfactant for 2 min, placed in 2.5% commercial bleach for 2 min, and then rinsed three times in sterile distilled water. Up to three leaf sections, 2–3 mm in width, are removed from the basal end of the leaf and plated on solid SH-30 medium. Following culture initiation, callus tissue will form in 10 to 15 days. After 2 to 3 weeks of culture, localized areas of white, embryogenic callus appear with many proembryos on the surface of the callus. Three to four weeks following culture initiation, numerous embryos at different stages of development (globular to scutellar stages) are found over the surface of the callus. If maintained for an extended period on SH-30 medium, (5 weeks or more), the embryos will enlarge, dedifferentiate, and produce further embryogenic callus (Tar'an, 1992).

For large-scale embryo production, callus is transferred after 2–3 weeks on solid SH-30 medium to liquid SH medium (1 g/100 mL) containing 30 μ*M* dicamba and casein hydrolysate. Every 3 weeks, the cell suspension is filtered through successive 500 and 224 μm Nitex® mesh sieves and the 224 μm fraction either transferred to fresh induction medium (1 g/100 mL) to maintain the suspension culture or transferred to hormone-free liquid SH medium (1 g/100 mL) for embryo development. Following 1 week in development medium, the cell suspension is filtered through successive 1000 and 100 μm Nitex® mesh sieves and the 100 μm fraction transferred to solid hormone-free SH medium for embryo germination.

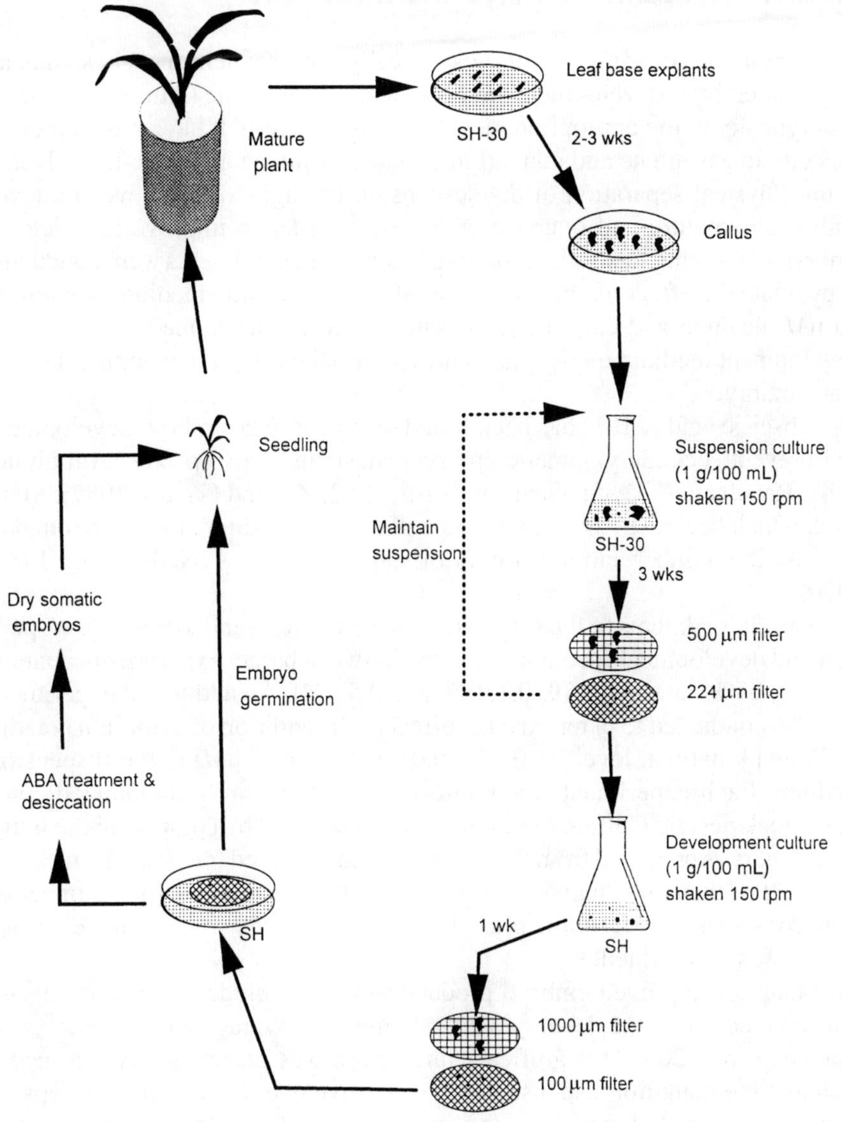

Figure 13.1. Outline of the cell culture system to produce somatic embryos in orchardgrass. Callus, suspension, and development cultures are maintained in the dark at 25°C, while embryo desiccation and germination phases are in the light.

EMBRYO YIELD AND SYNCHRONIZATION

Populations of somatic embryos typically form asynchronously so that all stages of embryo development are present at any one time (Ammirato, 1983). Embryos not at the appropriate developmental stage will have reduced or no capacity to germinate and convert into viable plants on hormone-free SH medium. Physical separation of the suspension through Nitex® sieves removes both the immature and oversize embryos. In order to improve the yield of embryos following mechanical sieving, a series of experiments were conducted to evaluate the effect of changes to the SH-30 suspension medium containing 30 μ*M* dicamba and casein hydrolysate and to the hormone-free liquid SH development medium on the yield and synchronization of development of somatic embryos.

Abscisic acid (ABA) has been found to synchronize embryo development and prevent secondary somatic embryogenesis in many species (Ammirato, 1987; Ranch et al., 1985; Vasil and Vasil, 1982; Ziv and Gadasi, 1987). Mannitol, which decreases the osmotic potential of the medium, has been found to increase the number and uniformity of somatic embryos (Nadel et al., 1989, 1990).

Two amendments to the suspension, development, and to both the suspension and development media were tested in two separate experiments: mannitol (0, 3,4,5%), and ABA (0, 0.1, 0.3, and 0.5 μ*M*). An additional experiment was also conducted to determine the effect of the addition of cytokinin (zeatin, BAP, and kinetin at levels 0, 0.001, 0.01, 0.1, or 1.00 μ*M*) to the suspension medium. Each experiment was replicated four times in a randomized complete block design. Culture treatments were evaluated by comparing the number of embryos produced from the E1 (0.1–1.0 mm) sieved fractions. To measure the effect of the amendments on embryo synchronization, the total number of embryos — the E1 fraction plus the larger embryos lost in the E2 fraction — was determined.

Mannitol improved embryo production when included in either the suspension medium or the development medium. The average increase across all treatments was 20%. A significant interaction was detected between media such that the mannitol response was not additive over the two media steps. If mannitol was included in both the suspension and development media, the increase in embryo number (average 16%) was similar to the increase obtained if mannitol was only added to the development medium (average 16%). (See Table 13.1.)

The addition of either cytokinin (BAP, zeatin, or kinetin) or ABA significantly decreased the number of E1 fraction embryos. The decline in embryo numbers averaged 33% over all ABA treatments. The decline in embryo numbers averaged 73% over all cytokinin levels. (Data not presented.)

The addition of mannitol to only the suspension medium reduced the fraction of total embryos collected in the E1 fraction. If mannitol was also included in the development medium, this detrimental effect was negated such

Table 13.1. Effect of Mannitol and ABA on Somatic Embryo Production of Orchardgrass Measured by the Number of E1 (0.1–1.0 mm) Somatic Embryos as a Percentage of the Control and the Number of E1 Embryos as a Percentage of Total Embryo Production.[a]

Treatment		Number of E1 Fraction Embryos			Number of E1 Embryos		
		Suspension	Development	Both	Suspension	Development	Both
			% of control			% of total	
Mannitol	0%	100 b	100 b	100 b	97.9 a	97.9 a	97.9 a
	3%	105 b	114 ab	103 b	95.0 b	98.3 a	96.8 a
	4%	150 a	127 a	119 a	92.1 c	98.9 a	97.6 a
	5%	114 b	107 b	126 a	90.0 c	99.8 a	95.9 a
ABA	0.0 μ*M*	100 a	100 a	100 a	71.6 b	71.6 b	71.6 b
	0.1 μ*M*	61 bc	85 ab	63 b	75.4 ab	91.6 a	78.3 ab
	0.3 μ*M*	66 b	74 b	71 b	76.1 ab	90.0 a	79.5 a
	0.5 μ*M*	41 c	77 ab	66 b	80.3 a	85.3 a	80.7 a

[a] Means are an average of four replications. Means within a column within a treatment followed by the same letter are not significantly different using a Duncan's multiple range test (μ=0.05).

that nearly all embryos (98%) were retained within the E1 fraction (Table 13.1). The addition of 0.1 μ*M* ABA also increased the fraction of embryos collected in the E1 fraction (Table 13.1), as well as reducing the number of embryos that precociously germinated from approximately 1% of total embryos to 0.1% of total embryos (data not presented).

These studies to date indicate that 3–4% mannitol should be included in the suspension and development media in order to increase the total number and fraction of E1 embryos. Although ABA reduced the total number of embryos, the addition of 0.1 μ*M* ABA to the development medium increased the fraction of E1 embryos obtained and reduced precocious germination.

INHERITANCE OF SOMATIC EMBRYOGENESIS

The ability of plant tissue to regenerate following *in vitro* culture is highly dependent on the genotype (Hanning and Conger, 1986; Kielly and Bowley, 1992). The inheritance of somatic embryogenesis in orchardgrass was studied using one embryogenic genotype (Regen-O) and random plants from two non-embryogenic populations, Jay and OSG-13. Individual plants from F_1 (Regen-O X Jay), F_2, BC_1, and test cross (F_1 X OSG-13) matings were evaluated for somatic embryogenesis.

Reciprocal F_1 crosses, without emasculation, were made between Regen-O and ten random plants from the variety Jay to produce ten F_1 families. Approximately 20 progenies of each F_1 family were screened for somatic embryogenesis. Crosses between selected F_1 plants were then made to produce a series of Embryogenic F_1 X Embryogenic F_1 (E X E) and a series of Non-embryogenic F_1 X Non-embryogenic F_1 (Ne X Ne) matings. Sets of backcrosses were also produced between embryogenic F_1 plants and an additional random set of plants from the non-embryogenic variety Jay (Embryogenic F_1 X Jay). In addition, test crosses were also produced between selected embryogenic F_1 plants and random plants from the non-embryogenic population, OSG-13.

The ability to form somatic embryos from cultures derived from leaf base explants was under nuclear control; there was no evidence of cytoplasmic effects. The segregation ratio observed across all matings indicated that somatic embryogenesis in orchardgrass was conditioned by two independent dominant genes having complementary effects. For example, the results of the progeny of embryogenic F_1 plants backcrossed to the non-embryogenic parent (Jay) and of embryogenic F_1 plants test crossed to another non-embryogenic population (OSG-13) are summarized in Table 13.2. Four possible genetic models — a one-locus and three two-locus models — were tested. The pooled segregation ratio of the BC_1 (F_1 X Jay) and of the test cross to OSG-13 (F_1 X OSG-13) corresponded to the expected ratio for a cross between simplex/simplex (AaaaBbbb) and nulliplex/nulliplex (aaaabbbb) genotypes.

Table 13.2. Chi-square (χ^2) Goodness of Fit Values Comparing Observed and Expected Embryogenic (E) and Non-Embryogenic (Ne) Segregation Ratios, Assuming a One- or Two-Locus Dominant Tetraploid Model, for BC_1 Progeny of Orchardgrass Derived from Crosses Between Embryogenic F_1 Progeny and Ne Plants from the Populations Jay and OSG-13.[a]

		Model Tested			
		1-Locus	2-Locus		
		Aaaa x aaaa	AaaaBbbb x aaaabbbb	AaaaBbbb x Aaaabbbb	AAaaBbbb x aaaabbbb
		Expected Ratio (E:Ne)			
Cross	**Observed Ratio (E:Ne)**	**1:1**	**1:3**	**3:5**	**5:7**
	no.	χ^2			
F_1 x Jay	16:52	18.01**[b]	0.02 NS	5.08*	8.47**
Jay x F_1	20:50	12.01**	0.30 NS	2.02 NS	4.41*
Combined	36:102	30.62**	0.04 NS	7.19**	13.15**
F_1 x OSG-13	28:53	7.11**	3.46 NS	0.19 NS	1.40 NS
OSG-13 x F_1	19:65	24.11**	0.14 NS	7.31**	11.77**
Combined	47:118	29.70**	0.89 NS	5.34*	11.26**

[a] Observed ratios are pooled across five Jay x F_1 and three OSG-13 x F_1 families.

[b] **, * = Significant at 1% and 5% levels, respectively; NS = nonsignificant.

The simple model of inheritance and the dominance of the character simplifies selection for populations with improved *in vitro* response in orchardgrass. Back-cross breeding and recurrent selection has been employed successfully to increase frequency of regenerative plants in many species, including *Medicago sativa* L. (Bowley et al., 1993), *Trifolium pratense* L. (Quesenberry and Smith, 1993), *Lycopersicon esculentum* Mill. (Koornneef et al., 1987), *Zea mays* L. (Petolino et al., 1988; Mórocz et al., 1990; Armstrong et al., 1992) and *Brassica* (Aslam et al., 1990). Transfer of the somatic embryogenesis trait to non-regenerative populations, which otherwise have superior agronomic characters, will enable development of commercial orchardgrass varieties with a high regeneration capacity. Combining the ability to form somatic embryos with agronomic utility will ease the application of biotechnological techniques in orchardgrass.

REFERENCES

Ammirato, P.V. 1983. Embryogenesis, in *Handbook of Plant Cell Culture,* Vol. 1. Evans, D.A., W.R. Sharp, P.B. Ammirato, and Y. Yamada, Eds., Macmillan Publ. Co., NY.

Ammirato, P.V. 1987. Organizational Events During Somatic Embryogenesis, in *Plant Tissue and Cell Culture*, Green, C.E., D.A. Somers, W.P. Hackett, and D.D. Biesboer, Eds., Alan R. Liss, Inc., NY.

Armstrong, C.L., J. Romero-Severson, and T.K. Hodges. 1992. Improved tissue culture response of an elite maize inbred through backcross breeding, and identification of chromosomal regions important for regeneration by RFLP analysis. *Theoretical and Applied Genetics,* 84:755–762.

Aslam, F.N., M.V. MacDonald, P.Loudon, and D.S. Ingram. 1990. Rapid-cycling *Brassica* species: Inbreeding and selection of *B. campestris* for anther culture ability. *Annals of Botany,* 65:557–566.

Bowley, S.R., G.A. Kielly, K. Anandarajah, B.D. McKersie, and T. Senaratna. 1993. Field evaluation following two cycles of backcross transfer of somatic embryogenesis to commercial alfalfa germplasm. *Canadian Journal of Plant Science,* 73:131–137.

Conger, B.V. and G.E. Hanning. 1991. Registration of Embryogen-P orchardgrass germplasm with a high capacity for somatic embryogenesis from *in vitro* cultures. *Crop Science,* 31:855–856.

Hanning, G.E. and B.V. Conger. 1986. Factors influencing somatic embryogenesis from cultured leaf segments of *Dactylis glomerata* L. *Journal of Plant Physiology,* 123:23–29.

Kielly, G.A. and S.R. Bowley. 1992. Genetic control of somatic embryogenesis in alfalfa. *Genome,* 35:474–477.

Koornneef, M., C.J. Hanhart, and L. Martinelli. 1987. A genetic analysis of cell culture trait in tomato. *Theoretical and Applied Genetics,* 74:633–641.

McKersie, B.D. and S.R. Bowley. 1993. Synthetic Seeds of Alfalfa, in Redenbaugh, K., Ed., *Synseeds: Applications of Synthetic Seeds to Crop Improvement*, CRC Press, Boca Raton, FL.

Mórocz, S., G. Donn, J. Nemeth, and D. Dudits. 1990. An improved system to obtain fertile regenerants via maize protoplasts isolated from a highly embryogenic suspension culture. *Theoretical and Applied Genetics,* 80:721–726.

Nadel, B.L., A. Altman, and M. Ziv. 1989. Regulation of somatic embryogenesis in celery cell suspension: II. Early detection of embryogenic potential and the induction of synchronous cell cultures. *Plant Cell Tissue and Organ Culture,* 20:119–124.

Nadel, B.L., A. Altman, and M. Ziv. 1990. Regulation of large scale embryogenesis in celery. *Acta Hort.,* 280:75–85.

Petolino, J.F., A.M. Jones, and S.A. Thompson. 1988. Selection for increased anther culture response in maize. *Theoretical and Applied Genetics,* 76:157–159.

Quesenberry, K.H. and R.R. Smith. 1993. Recurrent selection for plant regeneration from red clover tissue culture. *Crop Science,* 33:585–589.

Ranch, J.P., L. Oglesby, and A.C. Zielinski. 1985. Plant regeneration from embryo derived tissue cultures of soybeans. *In Vitro Cell Dev. Biology,* 21:653–658.

Tar'an, B. 1992. "Somatic embryogenesis in orchardgrass (*Dactylis glomerata* L.)," thesis presented to the University of Guelph, Guelph, Ontario, in partial fulfillment of the requirements for the degree of Master of Science.

Vasil, V. and I.K. Vasil. 1982. Characterization of an embryogenic cell suspension culture derived from cultured inflorescence of *Pennesitum americanum* (pearl millet, Graminae). *American Journal of Botany,* 69:1411–1449.

Ziv, M. and G. Gadasi. 1987. Enhanced embryogenesis and plant regeneration from cucumber (*Cucumis sativus*) callus by activated charcoal in solid/liquid double-layer cultures. *Plant Science,* 74:115–122.

Chapter 14

Plant Breeding, Plant Regeneration, and Flow Cytometry in Buffalograss

T.P. Riordan, S. Fei, P.G. Johnson, and P.E. Read

Buffalograss (*Buchloë dactyloides* [Nutt.] Engelm.) is a native grass species that has only recently been considered as an alternative turfgrass species. (Engelke and Lehman, 1990; Huff and Wu, 1987; Riordan et al., 1983). However, it is not new, but only rediscovered due to concerns about water and the environment. Buffalograss has been used for lawns in rural areas of the central Great Plains since at least the 1880s (Beetle, 1950; Wenger, 1943). It was even recommended for golf course fairway use in early United States Golf Association (USGA) publications. Buffalograss was one of the few species that persisted during the dust bowl in the 1930s (Beetle, 1950; Wenger, 1943).

Buffalograss is possibly the environmentally correct species for turfgrass use in the next century. It already has characteristics that are being sought in other conventional turfgrass species. Buffalograss is considered a low user of water and energy. It also has low requirements for nitrogen, pesticides, and mowing. It probably is not a "no maintenance" species, but it can easily be managed as a low maintenance species.

Although buffalograss is considered a species for low maintenance, some use may be at the higher maintenance levels of home lawns and golf courses. Research conducted at Nebraska indicates that buffalograss performs well, if not better, at higher levels of management. Turfgrass quality of some genotypes may be higher at low mowing heights (below 2.0 cm) and at increased application levels of nitrogen (above 10 g N/m^2/growing season) and water. It is even possible that some genotypes are better adapted to this closer mowing height, which does not seem to significantly increase nitrogen or water requirements. These attributes may be very significant for future use on golf courses.

The USGA has supported the buffalograss research project at Nebraska since 1984, and this support has led to the release of five cultivars and the planned release of three new cultivars in 1996. These releases have been well accepted in the marketplace, especially in areas where there are water shortages or other significant environmental concerns. However, it is possible that buffalograss will be only another turfgrass alternative and will be used on its own merits, including quality, cold tolerance, and a low requirement for water and other inputs.

Even if buffalograss had no positive turfgrass characteristics, it would probably still be an excellent plant to study, because it is obviously different from other turfgrass species. The main differences are as follows: It is a warm-season species that has excellent cold tolerance. It is a dioecious species, having both male and female plants, and occasionally with both sexes occurring separately on the same plant. And finally, buffalograss also has different ploidy levels and associated areas of adaptation and turfgrass characteristics (anatomical and growth habit). The diploid plants (2n=2x=20) are found in Mexico and parts of southern Texas and have, in general, poor cold hardiness and sod strength. Buffalawn is an example of this type. The tetraploids (2n=4x=40) are found in Central Texas and have moderate cold tolerance, and at least some have good sod strength. Prairie and 609 are examples of tetraploid buffalograsses (Engelke et al., 1990; Riordan et al., 1992). The hexaploid (2n=6x=60) buffalograsses are found in the northern Great Plains and have excellent cold hardiness, are pubescent, and have poor to moderate sod strength. Examples of hexaploid buffalograsses include the cultivars 315 and 378 (Riordan et al., 1995).

These ploidy levels may allow buffalograss to have a wide range of adaptation compared to other warm-season species, with types found from Mexico to Canada. Although it is not known why various ploidy levels are adapted to certain areas of the Great Plains, it has been possible to manipulate this adaptation to the breeder's advantage. Southern adapted genotypes, such as 609, have average cold hardiness, and when planted in the northern portion of the Great Plains, they have an extended growing season but with reduced cold hardiness. The northern adapted genotypes, such as 315, have excellent cold hardiness and can be planted both in the North and the South, but with a reduced growing season.

It has also been observed that Prairie and 609, both tetraploid buffalograsses, are non-pubescent and are highly tolerant to two species of the buffalograss mealybug [*Tridicus sporoboli* (Cockerell) and *Trionymus* spp.] (Baxendale et al., 1994; Johnson-Cicalese, 1995). Although it seems that the lack of pubescence is a non-preference characteristic to the mealybug, there are probably other factors that also affect tolerance.

In our ten years of research, it has been quite obvious to each of us that we have a lot to learn about buffalograss as a *new* turfgrass species. We have learned that:

1. It is a warm-season species that can have excellent cold tolerance.
2. It is a low maintenance species that can he maintained at a much higher level.
3. The ploidy level is related to adaptation and various plant and growth characteristics.

It is therefore important that we identify the ploidy level of the genotypes we are trying to improve.

FLOW CYTOMETRY

Introduction

With the wide variation in chromosome number in buffalograss, it is essential to determine the ploidy level of accessions in breeding projects to improve hybridization and population development. Crossing plants of different ploidy levels may produce progeny with an uneven number of genomes, such as triploids and pentaploids, which are often infertile (Poehlman, 1987) or the seed from them may exhibit undesirable variation. On the other hand, the uneven ploidy levels, or individuals with intermediate chromosome numbers, may have unique characteristics that can be utilized in a vegetatively propagated variety (Fehr, 1987: Chapter 4), like buffalograss, bermudagrass, and zoysiagrass.

Unfortunately there are no dependable morphological markers that can determine ploidy level in buffalograss. And the small chromosome size of all buffalograsses, and large numbers of chromosomes in the tetraploids and hexaploids, make this evaluation difficult with the usual cytological techniques. With these methods, usually only one to two accessions can be studied per day. We have used flow cytometry to successfully estimate chromosome number of buffalograsses quickly and easily. Compared to the cytological techniques, well over 100 ploidy level determinations can be made per day with flow cytometry. Up to this date, flow cytometry has been used to a limited extent in turfgrass breeding, with one reference on *Poa pratensis* (Huff and Bara, 1993).

In this section, we will describe our methods and results from our DNA content evaluations of buffalograss, and consider applications of flow cytometry for further characterization of buffalograss and other turfgrass species.

Technique Description

We briefly describe the flow cytometry methods in measuring nuclear DNA content. For readers interested in more detailed descriptions of theories, techniques, and equipment, we recommend reading Galbraith et al. (1983), Arumuganathan and Earle (1991), and Givan (1992).

For our use, flow cytometry quantifies the DNA content of cell nuclei by measuring the fluorescence of DNA-specific dye as it is excited by monochromatic light (laser). The first step is to prepare the plant tissue to isolate cell nuclei (nuclear DNA quantification). For buffalograss we use the protocol outlined by Arumuganathan and Earle (1991). This method consists of excising 30 mg of leaf tissue and chopping it into narrow strips while in a solution consisting of the fluorochrome propidium iodide (PI) and a buffer solution that lyses the cells to release intact nuclei. PI intercalates within the DNA and RNA molecules. The slurry is then filtered with 33 μm nylon mesh, centrifuged to pellet the nuclei, resuspended in the extraction solution along with DNAse-free RNAse to degrade any RNA present in the sample, and incubated. Once the sample is prepared, it is analyzed at the University of Nebraska Flow Cytometry Core Research Facilities using a Becton Dickinson FACScan flow cytometer. The PI-stained nuclei fluoresce when excited with a laser with an intensity directly correlated to the amount of DNA in the nuclei. Corn leaves were chopped and prepared in the same sample, along with the buffalograss tissue, as an internal standard used for calculating the genome size of the unknown buffalograss accession. The size of the corn genome is 5.3 pg/nucleus. The buffalograss and corn in the sample preparation create distinct peaks of fluorescence intensity (Figure 14.1). The nuclear DNA content is estimated by comparing the position of the peak created by the plant with the unknown DNA content — (buffalograss) with the peak created by the internal standard (corn) — as shown below:

$$\text{nuclear DNA amount of unknown tissue} = \frac{\text{mean of unknown peak}}{\text{mean of internal standard peak}} \times \text{DNA content of internal standard}$$

Chromosome counts have been made on root-tip meristems of a number of accessions using the Fuelgen staining method.

Results and Discussion

In the measurements we have done thus far, it is obvious that buffalograss shows significant variation in DNA content (Figure 14.2). Exact chromosome counts have been made on the diploid (20 chromosome) and tetraploid (40 chromosome) accessions, and approximate counts have been made on the hexaploid accessions. The chromosome counts fit very closely with the three groupings in the DNA content measurements. The means of the three groups are 0.93, 1.83, and 2.59 pg/nucleus for the diploids, tetraploids, and hexaploids, respectively, with little variation within each group. However, 315 is

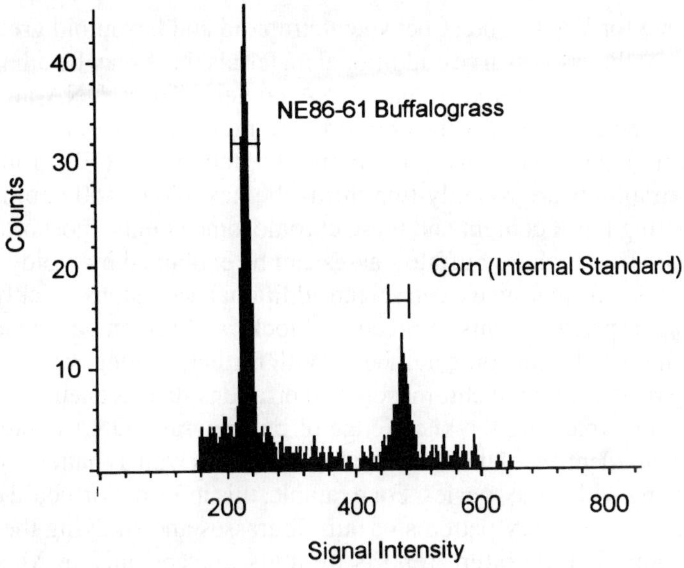

Figure 14.1. Analysis of NE86-61 buffalograss nuclei by flow cytometry. Histogram shows peaks of signal intensity (fluorescence) of buffalograss and corn nuclei. Signal intensity is a linear scale. Buffalograss peak has a mean at 224.40 and corn at 459.20.

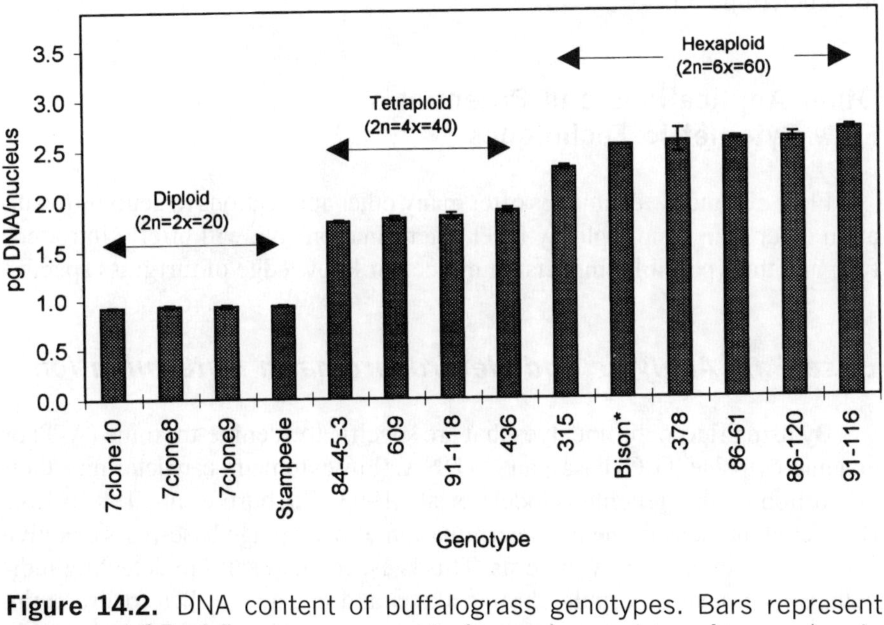

Figure 14.2. DNA content of buffalograss genotypes. Bars represent mean and SE of five measurements for each genotype. Arrows denote ploidy levels determined by chromosome counts.

intermediate for DNA content between tetraploid and hexaploid groups. This may indicate that as we survey additional materials there may be other factors, other than ploidy level, that impact DNA content. These DNA amounts fit what would be expected for the chromosome numbers where the diploid (20 chromosome) genotypes have roughly half the tetraploids (40 chromosome), and the tetraploids are roughly two-thirds the hexaploids (60 chromosome). By correlating DNA content and these chromosome counts, most of the DNA content variation in these buffalograsses can be explained by ploidy level.

With this information we can screen additional accessions quickly and put together appropriate parents in breeding blocks and screen additional accessions for unique chromosome numbers. With further refining, we may be able to identify relatively small chromosomal aberrations such as aneuploids, which may assist in expanding the knowledge of buffalograss. DNA content variation may help identify relatedness to other species as well as patterns of evolution occurring within the species. For example, this information could be useful in studying evolutionary patterns of buffalograsses and studying the patterns of adaptation. The literature reports diploids present only in Mexico and tetraploids along the western periphery of the Great Plains. Hexaploids are reported most prevalent throughout most of the range (Huff et al., 1993). However, others suggest tetraploids evolved from diploids, thereby extending the species' range into the southern Great Plains. Further development of the hexaploids may have extended the range of adaptation further north to what we are familiar with today (Reeder and Reeder, 1972). The rapid screening technique would speed up this kind of evaluation dramatically, with more reliable ploidy level determinations.

Other Applications and Potential of Flow Cytometric Techniques

Flow cytometry techniques offer many other applications to genome evaluation other than simple ploidy level determination. We will briefly introduce two, and their possible impacts on molecular knowledge of turfgrass species.

Base-Pair Analysis and Heterochromatin Determination

By using fluorochrome dyes that are specific to adenine-thymine (A-T) or guanine-cytosine (G-C) base-pairs in DNA, flow cytometry can determine their proportion of the genome (Godele et al., 1993; Rayburn et al., 1992). Like DNA content measurements, determination of percentage base-pairs can give evidence for evolutionary patterns. This is especially useful in detecting individuals with heteromorphic chromosomes and changes in heterochromatin, since heterochromatin is A-T rich. Godele et al. (1993) suggest that change in heterochromatin is rapid, relative to biological evolution, so closely related species should have similar DNA contents and similar base-pair composition. Initial efforts to compare buffalograss base-pair ratios showed no difference among ploidy levels or gender (Johnson et al., 1996).

Chromosome Sorting

Flow sorting of chromosomes is an exciting technology that permits the isolation of relatively large quantities of single chromosomes. This sorted chromosomal DNA is ideal for gene-mapping and the construction of chromosome-specific DNA libraries, rather than total genome libraries. In fact, this type of technology underlies much of the human genome project by speeding up the mapping to specific chromosomes (Givan, 1992; Dolezel et al., 1994). Having isolated chromosomal DNA of turfgrass species would likely be helpful to most of the topics being presented at this workshop.

Unfortunately, the use of this technology on plants has been slow, since many species have morphologically similar chromosomes, which makes sorting by size difficult (Dolezel et al., 1994). Buffalograss is no exception. However, several plant species have been sorted, including: *Petunia hybrida*, *Vica faba*, *Zea mays*, and *Triticum aestivum* (Dolezel et al., 1994). As the technology is rapidly refined (Givan, 1992), flow sorting may be applicable to turfgrass species in the near future.

IN VITRO REGENERATION OF BUFFALOGRASS

Introduction

Although buffalograss is a dioecious species, it is possible to use the conventional breeding techniques of an open-pollinated species. The major disadvantage of these conventional breeding techniques is the time required to make genetic improvement. The recent development of plant transformation technology using particle bombardment may provide for more immediate prospects for improving buffalograss. In addition, the use of somaclonal variation may allow for the creation and screening of desirable variants of buffalograss *in vitro*. In order to achieve these goals, a highly reproducible regeneration procedure is required.

Regeneration Methods

Immature inflorescences from two female clones, 609 and 315, and a male clone, NE84-45-3, were harvested from field-grown plants. The inflorescences were disinfected for five minutes in 70% ethanol, followed by three rinses in sterilized distilled water before placing them on the tissue culture media. Mature caryopses of 'Cody' were disinfected by using undiluted commercial bleach for 30 min followed by three rinses of sterilized distilled water. The basic culture medium was as described by Murashige and Skoog (1962). Various concentrations of 2,4-D, NAA, BA, and combinations of the three, were used in the callus induction media. The regeneration media contained 0.0 to 0.5 mg BA/L. All cultures were kept in dark conditions at 25°C.

Callus Induction Results

Compact, light yellow calli with nodular structures were formed in all 2,4-D treatments, ranging from 2 to 10 mg/L after one month in the dark. The optimum concentration was 2 mg/L. NE84-45-3 was the most responsive genotype, with about 50% calli formation. Callus also formed on NAA-containing media ranging from 2 to 10 mg/L, with the optimum at 8 mg/L NAA. However, nodular structures were formed only on media containing more than 2 mg/L NAA. The size of callus masses formed on NAA media was much smaller than those formed on the 2,4-D-containing media. Quantity of callus was increased when these cultures were extended beyond one month or calli were transferred to fresh media. When callus with nodular structures is isolated from non-embryogenic background callus, the production of a mucilage-like substance is increased. Unfortunately, this substance has been correlated with a decrease or total lack of embryogenic potential (Hanning and Conger, 1982). As for the caryopsis culture, whitish and watery calli without any organized structures were observed after one month of culture on treatments containing either 2,4-D alone (ranging from 2 to 10 mg/L) or with 2,4-D combined with BA (ranging from 0.1 to 0.4 mg/L). Multiple shoots were observed on medium containing 4 mg, 2,4-D/L, and 0.1 mg BA/L. However, the frequency was only about 3%. Subculture of these whitish and watery callus masses occasionally led to development of compact nodular structures similar to those developed through immature inflorescence cultures.

Plant Regeneration Results

After transferring organized calli onto a regeneration medium, a noticeable difference in regeneration pattern was found between calli derived from NAA treatments and 2,4-D treatments. Calli developed on NAA-containing media produced intact plants with normal shoots and roots about 20 days after they were transferred to regeneration medium without plant growth regulators. Calli that developed on 2,4-D-containing media produced only shoots when they were transferred to regeneration medium supplemented with 0.1–0.5 mg BA/L. Another transfer to a rooting medium that contained half-strength MS medium and 0.3 mg NAA/L was necessary to obtain root formation. Since the calli formed on NAA treatments were small in size and developed at a very low frequency, the overall regeneration frequency was much lower than was found for 2,4-D treatments.

Future Prospects

To fully make use of the particle bombardment technology, a regeneration protocol with high frequency is critical, since only a small amount of cells will be transformed. Currently, the regeneration frequency for particularly desirable cultivars is still low. For example, only about 17% of the callus masses with nodular structure from '609' produced shoots. In order to increase the regeneration frequency, various efforts have been initiated, including: (1) modification of medium components for callus induction and regeneration; (2) screening for responsive genotypes and explant types; and (3) using suspension culture. We have found that the male clone NE84-45-3 is more responsive than the two female clones, '609' and '315.' Using suspension culture techniques, we have obtained somatic embryos from '609' and NE84-45-3. However, maturation and germination have not yet been achieved.

Another research area we are working on is haploid production through anther culture. Buffalograss is a dioecious species, and although hermaphrodite flowers occur naturally at a low frequency, they are reported as self-incompatible (Huff et al., 1993). By doubling the chromosomes of haploids, a high degree of homozygosity may be quickly achieved. These lines can then be used for improvement of seeding vigor and uniformity by hybrid production. Presently, we have cultured anthers from NE84-45-3 and 'Texoka.' Cytological examination revealed that microspores of NE84-45-3 have started androgenesis and multicellular structures have been observed. Further study is under way.

CONCLUSION

Buffalograss is a native grass species that shows excellent potential for new uses as a turfgrass in many parts of the country. Its wide adaptation allows its culture throughout the Great Plains from Mexico to Canada, as well as in the desert Southwest and the Transition Zone. It has low requirements for water, fertilizer, pesticides, and mowing, and in general could be a turfgrass that has minimal negative affect on the environment. Breeding efforts with this species have led to the development and release of five cultivars and to potential additional improvements in the future. Additional studies will be required relating to the dioecious nature of buffalograss, the different ploidy levels and their effect on adaptation, and the basic knowledge of the plant. Much of these studies will involve work at the molecular level and will use the new techniques for genetic manipulation.

ACKNOWLEDGMENTS

We wish to thank K. Arumuganathan and his group for their assistance at the Flow Cytometry Core Research Facility at the University of Nebraska–Lincoln.

REFERENCES

Arumuganathan, K. and E.D. Earle. 1991. Estimation of Nuclear DNA Contents of Plants by Flow Cytometry. *Plant Molecular Biology Reporter*. 9:229–233.

Baxendale, F.P., J.M. Johnson-Cicalese, and T.P. Riordan. 1994. *Tridiscus sporoboli* and *Trionymus* sp. (Homoptera: Pseudococcidae): Potential New Mealybug Pests of Buffalograss Turf. *Journal Kansas Entomological Society*. 67(2):169–172.

Beetle, A.A. 1950. Buffalograss — Native of the Shortgrass Plains. *Agr. Exp. Sta., Univ. of Wyoming (Laramie) Bulletin*. 293:1–31.

Dolezel, J., S. Lucretti, and I. Schubert. 1994. Plant Chromosome Analysis and Sorting by Flow Cytometry. *Critical Reviews in Plant Sciences*. 13:275–309.

Engelke, M.C. and V.G. Lehman. 1990. Registration of 'Prairie' Buffalograss. *Crop Science*. 30:1360.

Fehr, W.R. 1987. Principles of Cultivar Development. In *Theory and Technique*, Vol. 1. McGraw-Hill. New York, NY.

Galbraith, D.W., K.R. Harkins, J.M. Maddox, N.M Ayres, D.P. Sharma, and E. Firoozbady. 1983. Rapid Flow Cytometric Analysis of Cell Cycle in Intact Plant Tissues. *Science*. 220:1049–1051.

Givan, A.L. 1992. *Flow Cytometry: First Principles*. Wiley-Liss, Inc., NY.

Godele, B., D. Cartier, D. Marie, S.C. Brown, and S. Siljak-Yakovlev. 1993. Heterochromatin Study Demonstrating the Non-Linearity of Fluorometry Useful for Calculating Genomic Base Composition. *Cytometry* 14:618–626.

Hanning, G.E. and B.V. Conger. 1982. Embryoid and plantlet formation from leaf segments of *Dactylis glomerata* L. *Theoret. Appl. Genet.* 63:159.

Haydu, Z and I.K. Vasil. 1981. Somatic Embryogenesis and Plant Regeneration from Leaf Tissues and Anthers of *Pennisetum purpureum* Schum. *Theoret. Appl. Genet.* 59:73.

Huff, D.R. and J.M. Bara. 1993. Determining Genetic Origins of Aberrant Progeny from Facultative Apomictic Kentucky Bluegrass Using a Combination of Flow Cytometry and Silver-Stained RAPD Markers. *Theoret. Appl. Genet.* 87:201–208.

Huff, D.R., R. Peakall, and P.E. Smouse. 1993. RAPD variation within and among natural populations of outcrossing buffalograss [*Buchloë dactyloides* (Nutt.) Engelm.]. *Theoret. Appl. Genet.* 86:927–934.

Huff, D.R. and Lin Wu. 1987. Sex expression in buffalograss under different environments. *Crop Science*. 27:623–626.

Johnson, P.G., T.P. Riordan, and K. Arumuganathan. 1996. Analysis of Buffalograss Germplasm Using Flow Cytometry. *Agronomy Abstracts*, p. 191.

Johnson-Cicalese, J.M. Evaluation of Buffalograss Leaf Pubescence and Its Effect on Mealybug Host Selection. University of Nebraska, *Turfgrass Research Report for 1994.*

Murashige, T. and F. Skoog. 1962. A revised medium for rapid growth and bioassays with tobacco tissue cultures. *Phys. Plant.* 15:473–497.

Ozias-Akins, P. and I.K. Vasil. 1982. Plant Regeneration from Cultured Immature Embryos and Inflorescences of *Triticum aestivum* L. (Wheat): Evidence for Somatic Embryogenesis. *Protoplasm.* 110:95–105.

Poehlman, J.M. 1987. *Breeding Field Crops, 3rd ed.* Van Nostrand Reinhold. NY.

Rayburn, A.L., J.A. Auger, and L.M. McMurphy. 1992. Estimating Percentage Constitutive Heterochromatin by Flow Cytometry. *Experimental Cell Research.* 198:175–178.

Reeder, J.R. and C.G. Reeder. 1972. Cytotaxonomy of *Buchloë dactyloides* (Graminaeae). *J. Colo.-Wyo. Acad. Sci.* 7:104–105.

Riordan, T.P., S.A. de Shazer, F. Baxendale, and M.C. Engelke. 1992. Registration of '609' Buffalograss. *Crop Science*. 32:1511.

Riordan, T.P., S.A. de Shazer, J.M. Johnson-Cicalese, and R.C Shearman. 1993. An Overview of Breeding and Development of Buffalograss. *International Turfgrass Society Research Journal.* 7:816–822.

Riordan, T.P., J. Johnson-Cicalese, F.P. Baxendale, M.C. Engelke, R.E. Gaussoin, G.L. Horst, and R.C. Shearman. 1995. Registration of '315' Buffalograss. *Crop Science*. 35:1206.

Wenger, L.E. 1943. Buffalograss. *Kansas Agri. Exp. Sta. (Manhattan) Bulletin.* 321:1–78.

Wu, L., A.H. Harvindi, and V.A. Gibeault. 1984. Observations on Buffalograss Sexual Characteristics and Potential for Seed Production Improvement. *HortScience*. 19(4):505–506.

Chapter 15

Herbicide-Resistant Transgenic Creeping Bentgrass

L. Lee and P. Day

INTRODUCTION

Weeds such as *Poa annua* in creeping bentgrass putting greens and fairways are a serious problem for golf course maintenance. Creeping bentgrass that is resistant to the herbicide glufosinate will allow the use of low application rates of herbicide to remove undesirable weeds with no effect on transgenic creeping bentgrass.

We are using genetic engineering technology for turfgrass improvement. A tissue culture and regeneration system is essential for gene transfer using presently available gene transfer techniques. At the AgBiotech Center, Rutgers University, we have successfully transformed creeping bentgrass through both biolistic transformation and protoplast transformation. In this chapter, we describe the development of transformation systems, field tests of transgenic plants, and the analysis of progeny from transgenic mother plants.

TISSUE CULTURE AND REGENERATION

To develop the transformation system, we first developed a tissue culture regeneration system for creeping bentgrass.

A total of 15 creeping bentgrass cultivars were used to initiate embryogenic callus cultures. All of them responded well. Depending on the cultivar, between 5 to 30% of surface sterilized mature seeds produced embryogenic callus cultures within 4–6 weeks after they were plated on a callus initiation medium. These callus cultures are highly regenerable. Upon transfer to regeneration medium, about 200–400 plantlets can be obtained from 1 g fresh weight of callus culture.

TRANSFORMATION

Two methods were developed for turfgrass transformation: biolistic transformation (Hartman et al., 1994) and direct DNA uptake into protoplasts (Lee et al., 1996). The *gus* gene was used as a scorable marker to optimize transformation parameters and the herbicide bialaphos was used for selection.

The herbicide-resistant gene, *bar*, encodes an enzyme PAT (phosphinothricin acetyltransferase) which inactivates the herbicide PPT (phosphinothricin) by acetylation (Murakami et al., 1986; Thompson et al., 1987). PPT is the active ingredient of glufosinate or bialaphos. Trade names of members of this group of herbicides are Basta®*, Herbiace®**, Ignite®*, Finale®*. The glufosinate group of herbicides inhibit glutamine synthetase and cause the accumulation of ammonia and cell death (Tachibana et al., 1986a,b). We used bialaphos for selection after transformation and Herbiace® or Finale® for herbicide treatments of regenerated plants.

Biolistic Transformation

A Bio-Rad PDS-1000/He biolistic delivery system was used to introduce DNA-coated gold particles to turfgrass culture cells from either suspension cell cultures or embryogenic callus cultures. The initial optimization using the transient expression of *gus* gene showed that gold particles produce 100x more blue foci than tungsten particles. Gold particles probably are less toxic to turfgrass culture cells.

Figure 15.1 is a flowchart of the biolistic transformation procedure. Bombarded cells were selected with 2 or 4 mg/L of bialaphos 3–7 days after bombardment, either with plate selection, or liquid then plate selection, for 8 weeks. Resistant colonies were further multiplied and plants were regenerated from them. In other experiments, bombarded materials were all regenerated and herbicide was applied to plants growing in soil in the greenhouse to screen for transformants.

Regenerated plants were transplanted to soil in a 96-well trays. One-month-old plants were treated with 2 mg/mL Herbiace®, at about 5x the field rate. No control plants survived herbicide treatment after 10 days, while transgenic

---

* Registered Trademark of AgrEvo Company, Inc., Wilmington, Delaware.

** Registered Trademark of Meiji Seika Kaishya, Ltd., Tokyo.

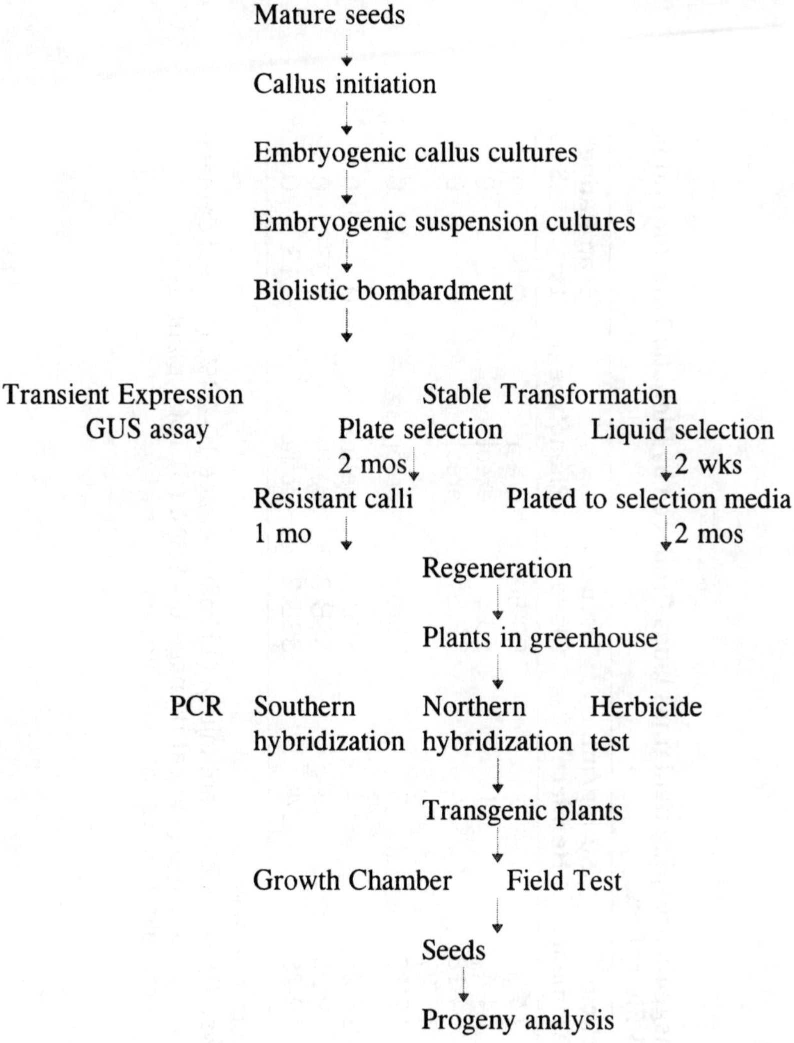

Figure 15.1. Flowchart of biolistic transformation of creeping bentgrass.

plants were green and healthy like untreated control plants. Table 15.1 shows that from five bombardment experiments, over 900 plants were treated with Herbiace®, and a total of 55 plants derived from cultivars Emerald and Southshore were completely resistant. PCR analysis showed the presence of the *bar* coding region in the transgenic plants but not in the control plants. Southern blot analysis confirmed the incorporation of the *bar* gene into the plant genome of transgenic plants. Northern blot analysis showed the expression of *bar* transcripts in transgenic plants but their absence in control plants.

Table 15.1. List of Transgenic Creeping Bentgrass Lines Produced by Biolistic Transformation that Survived the 1994 Field Test.

Cultivar	Particle Bombardment		2.0 mg/mL Herbiace®	Tissue[a] Clone	Phenotype	Mean Rating[b]	
						1×	3×
Emerald	1	14-3	3	EB.5	erect	0	0
	2	14-4	1	EB.5	erect	0	0
	3	14-18	14	EB.5	erect	0	0
	4	16-27	—				
	5	19-15	5	EBmm	creeping[c]	0	0
Southshore	1	16-24	4	SSB.2	petite	0	0
	2	19-25	3	SSB.2	petite	0.33	0.39
	3	19-28	4	SSB.2	petite	0.13	0.4

[a] Each clone derived from one seedling.
[b] Ratings were as follows: (0) No visible plant injury. (1) Leaf damage 10–24%. (2) Leaf damage 25–39%. (3) Leaf damage 40–59%. (4) Leaf damage 60–99%. (5) Total plant death.
[c] Creeping phenotype is equivalent to w.t. of two cultivars.

Protoplast Transformation

Protoplasts were isolated from embryogenic suspension cells and cultured using a feeder layer system (Rhodes et al., 1988). A thin layer of turfgrass suspension cells was placed on medium beneath a nitrocellulose membrane to serve as nurse cells to provide nutrients for protoplast regrowth to callus and eventually to plants on the upper surface of the membrane.

Figure 15.2 is a flowchart of protoplast transformation. For selection, the *bar* gene was introduced into protoplast using PEG (polyethylene glycol) to enhance the uptake. Bialaphos was used to select for resistant colonies 16 days after protoplast isolation and transformation. Resistant colonies were recovered from suspension cell cultures of several creeping bentgrass cultivars. Transgenic plants were regenerated from resistant Cobra colonies.

Transgenic plants survived the 2 mg/mL Herbiace® application in greenhouse treatment. Southern and Northern blot analyses confirmed incorporation and the presence of the transcript (Lee et al., 1996).

FIELD TESTS

During the summer of 1994, transgenic plants obtained from bombardment were field tested at Rutgers Research and Development Center in Bridgeton, New Jersey. A field test permit was obtained from USDA-APHIS, and vegetatively multiplied plants were set out in a random plot design. Transgenic plants derived from Emerald, Putter, and Southshore were tested for resistance to the herbicide Ignite® at 1x and 3x field rates with three replications. All clones that survived the 2 mg/mL Herbiace® greenhouse test were completely resistant to both rates, remaining green and healthy like untreated plants in the control plot. No control plants germinated from seeds and transplanted as seedlings survived herbicide application. All transgenic plants that survived the first application were unaffected by a second herbicide application.

Transgenic plants obtained from protoplast transformation were field tested in the summer of 1995. Again, all Cobra plants survived the 2 mg/mL Herbiace® greenhouse test and were resistant to 1x and 3x herbicide treatment in the field.

Ratings of herbicide treatments were scored two weeks after application using the following scale. 0 = completely resistant, 5 = dead. Table 15.1 summarizes the results of herbicide treatments from the 1994 field test. More than 30 Emerald and Southshore creeping bentgrass lines were resistant to the 3x field rate.

During the winter of 1994, transgenic plants from the field were vernalized in the field, as were some transgenic plants from protoplast transformation in a fenced area. These plants were transplanted to large pots and moved

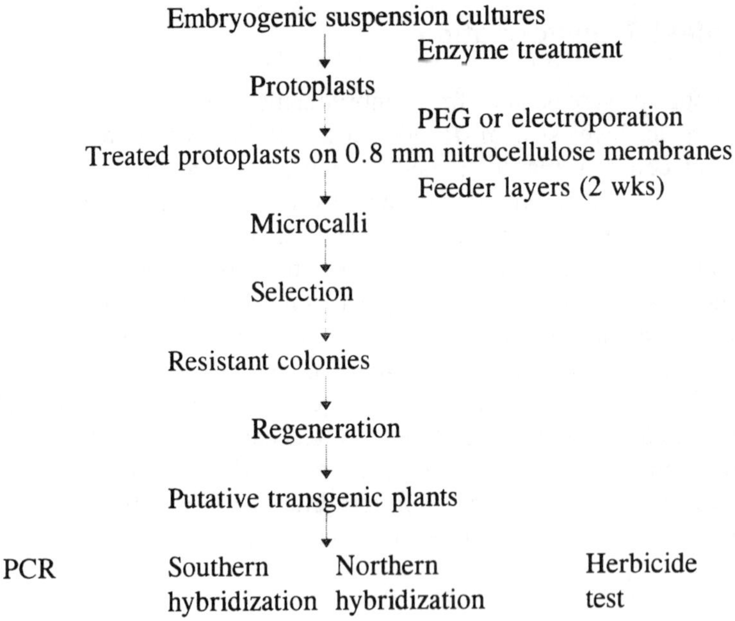

Figure 15.2. Protoplast regeneration and transformation scheme.

to a containment greenhouse before flowering. Open pollination occurred with control plants during the summer of 1995.

Progeny Analysis

Seeds were harvested from transgenic mother plants and control plants in the fall of 1995. Germination rates were comparable to control seeds used for callus initiation. Some seeds harvested from transgenic mother plants were able to germinate in the presence of herbicide, while control seeds were killed (Table 15.2).

After transplanting to soil, one-month-old seedlings were treated with 2 mg/mL Herbiace®. Some transgenic seedlings were resistant to the herbicide, with ratios in some lines about 1:1; and in other lines about 2:1. A very small number of seedlings germinated from control mother plants were also insensitive to herbicide. The result implies that transgenic mother plants produced viable pollen. Suitable resistant clones will be used as parents in a traditional breeding program before a resistant cultivar can be made available commercially.

With these herbicide-resistant transgenic plants, we can now study the outcrossing of transgenic creeping bentgrass to wild weedy plants. This problem will need to be addressed before the release of herbicide-resistant plants into the environment. Our success in the development of transformation sys-

Table 15.2. Germination Rates of Progeny of Herbicide-Resistant Transgenic Creeping Bentgrass Lines in Petri Dish or Greenhouse Herbicide Treatments.

Plant No.	Germination	Germination in Petri dish with Herbiace®			Greenhouse Test 2 mg/mL Herbiace®
		3x	1x	0.3x	
Cobra	86/100	0/100	0/100	0/100	0/29
Emerald	93/100	0/100	0/100	0/100	0/12
Southshore	82/100	0/100	0/100	0/100	na
585	10/17	na	na	na	5/9
955	41/50	na	50/50	na	na
SS1	44/50	na	na	na	3/36
4137	50/50	13/50	na	na	15/27
4143	70/100	na	na	na	11/24
4145	57/100	na	na	na	18/46
CR2	85/100	na	na	na	8/40
5061	99/100	19/50	39/50	48/50	34/53

tems for creeping bentgrass provides the selection tool for introducing other genes, such as disease resistance, into creeping bentgrass for future turfgrass improvement.

ACKNOWLEDGMENTS

We thank the United States Golf Association, the New Jersey Commission on Science and Technology, and the Turfgrass Center of Rutgers for their support of this work, and Drs. John Grande, Steve Johnston, and Brad Majak for assistance in the field trials.

REFERENCES

Hartman, C.L., L. Lee, P.R. Day, and N.E. Tumer. 1994. Herbicide resistant turfgrass (*Agrostis palustris* Huds.) by biolistic transformation. *Bio/Technology,* 12:919–923.

Lee, L., C.L. Laramore, P.R. Day, and N.E. Tumer. 1996. Transformation and regeneration of creeping bentgrass (*Agrostis palustris* Huds.) protoplasts. *Crop Sci.,* 36:401–406.

Murakami, T., H. Anzai, S. Imai, A. Satoh, K. Nagaoka, and C.J. Thompson. 1986. The bialaphos biosynthetic genes of *Streptomyces hygroscopicus*: Molecular cloning and characterization of the gene cluster. *Mol. Gen. Genet.,* 205:42–50.

Rhodes, C.A., K.S. Lowe, and K.L. Ruby. 1988. Plant regeneration from protoplasts isolated from embryogenic maize cell cultures. *Bio/Technology,* 6:56–60.

Tachibana, K., T. Watanabe, Y. Sekizawa, and R. Takematsu. 1986a. Inhibition of glutamine synthetase and quantitative changes of free amino acids in shoots of bialaphos-treated Japanese barnyard millet. *J. Pesticide Sci.,* 11:27–31.

Tachibana, K., T. Watanabe, Y. Sekizawa, and R. Takematsu. 1986b. Accumulation of ammonia in plants treated with bialaphos. *J. Pesticide Sci.,* 11:33–37.

Thompson, D.J., N.R. Movva, R. Tizard, R. Crameri, J.E. Davies, M. Lauwereys, and J. Botterman. 1987. Characterization of the herbicide-resistance gene *bar* from *Streptomyces hygroscopicus. EMBO J.,* 6:2519–2523.

Chapter 16

Simultaneous Control of Weeds, Dollar Spot, and Brown Patch Diseases in Transgenic Creeping Bentgrass

H. Zhong, C.-A. Liu, J.M. Vargas, Jr., D. Penner, and M.B. Sticklen

INTRODUCTION

Turfgrasses have been recognized for their importance to our quality of life for a long time and are cultured in nearly all inhabited regions of the world. Allotetraploid creeping bentgrass (*Agrostis palustris* Huds.) is one of the species of bentgrass that have been widely used for golf courses in the United States (Beard, 1982).

Weed infestation and disease infection are two of the major problems in golf courses and residential areas using creeping bentgrass. Due to the tolerance of grassy weeds to the herbicidal chemicals, the control of grassy weed infestation is very difficult in established creeping bentgrass greens and fairways. In addition, creeping bentgrass is susceptible to a variety of diseases, such as brown patch (*Rhizoctonia* spp.), dollar spot (*Sclerotinia homeocarpa*), *Pythium* blight (*Pythium* spp.), *Fusarium* blight (*Fusarium roseum* and *F. tricinctum*), and take-all patch (*Gaeumannomyces graminis*) (Smiley, 1983). Multiple applications of herbicides and pesticides to control weeds and diseases have become an indispensable tool of modern turf management. However, such efforts are consuming an increasing share of management expenses associated with growing concerns about the effects of these chemicals on the environment. Development of turfgrass varieties for better weed and disease control with less environmental pollution is of significant economic and social value.

Phosphinothricin (PPT), known as glufosinate, is the active ingredient of a nonselective, broad-spectrum, postemergent, commercial herbicide, Ignite®*. PPT is highly stable and rapidly degradable in a microbiologically active environment. PPT Contains a mixture of D-isomer, the inactive inhibitor of glutamine synthetase (GS), and the L-isomer, the active inhibitor of GS (Kondo et al., 1973; Hoerlein, 1994). In both plant and bacteria, it acts as a competitive inhibitor of GS (Bayer et al., 1972; Leason et al., 1982). GS is critical for the assimilation of ammonia as well as for general nitrogen metabolism. It is the only enzyme in plants that can detoxify ammonia released by nitrate reduction, amino acid catabolism, and photorespiration. The accumulation of ammonia caused the death of plant cells (Tachibana et al., 1986). Bialaphos, a precursor of L-PPT, is the active ingredient of the commercial herbicide Herbiace®**. Unlike PPT, bialaphos has better inhibitory activity against glutamine synthetase *in vitro,* but is converted to PPT *in vivo* by intracellular peptidases, which remove the alanine residues (De Block et al., 1987).

Resistance to PPT/bialaphos is conferred by phosphinothricin-N-acetyltransferase (PAT). PAT inactivates PPT by acetylation, using acetyl co-enzyme A as a cofactor. Two similar genes encode PAT: *bar*, isolated from *Streptomyces hygroscopicus* (Murakami et al., 1986; Thompson et al., 1987), and *pat*, isolated from *S. viridochromogenes* (Wohlleben et al., 1988).

Uchimaya et al. (1993) reported that transgenic rice plants expressing the *bar* gene could prevent infection of sheath blight pathogen (*Rhizoctonia solani*) after treatment of bialaphos. Since *R. solani* is also the etiologic agent of brown patch diseases of creeping bentgrass, the application of bialaphos to the bialaphos-resistant creeping bentgrass may also show reduction of the plant damage caused by fungal infection. Thus, a novel and economical usage of the herbicide would be to provide a simultaneous control of weeds and fungal pathogens in turf areas with bialaphos-resistant creeping bentgrass. Here, we report the first evidence on the prevention of the fungal infection in transgenic, bialaphos-resistant creeping bentgrass while using bialaphos to control weeds.

MATERIALS AND METHODS

Production of Transgenic Plants

Following the protocol described by Zhong et al. (1991), friable embryogenic callus was induced on callus induction medium, containing MS basal

---

* Registered Trademark of Hoecht-Roussel Agri-Vet company, Somerville, New Jersey.
** Registered Trademark of Meiji Seika Kaisha, Japan.

medium, 3% sucrose, 30 μ*M* 3,6-dichloro-*o*-anisic acid (dicamba) and 9 μ*M* 6-benzyladenine (BA), from caryopses of creeping bentgrass cultivar "Penncross." Embryogenic callus was selected and placed in a 2 cm^2 area on the surface of callus induction medium. After the plasmid pTW-a (Zhong et al., 1996) was precipitated onto tungsten particles following a published procedure (Zhong et al., 1993), the callus was bombarded three times with the precipitation mixture using the biolistic PDS-1000/He device. The bombarded calli were separated and cultured on the callus induction medium for three days. Then, they were subjected to selection on the callus induction medium, which contained 5 mg/L bialaphos or 15 mg/L PPT (ammonium salt), for 12 weeks at 4-week intervals. Plantlets regenerated from independent, resistant callus lines on MS basal medium containing the same amount of selective agent, and were transferred to soil in a greenhouse.

For confirmation of the presence and the expression of *bar* gene, the genomic DNA, RNA, and protein isolated from the leaf tissue of independently transformed plants in greenhouse were analyzed by Southern blot hybridization (Southern, 1975), Northern blot hybridization (Wadsworth et al., 1988) and thin-layer chromatography (D'Halluin et al., 1992).

Herbicide Application

The plants regenerated from independent transgenic callus lines were sprayed with 1.2% Ignite® (2.4 g/L of glufosinate ammonium, 150 L/ha). To determine the magnitude of their herbicide resistance, various rates of Ignite® (0 to 6%) were applied to three resistant lines. The concentrations of ammonia in transgenic plants of three lines after spraying with 1.2% Ignite® were evaluated.

In Vitro Responses of Pathogens to PPT and Bialaphos

Agar plugs with actively growing mycelium of *Rhizoctonia solani*, *Sclerotinia homeocarpa*, and *Pythium aphanidermatum* were cultured on the center of potato dextrose agar medium (PDA, 39 g/L; Difco Laboratories, Detroit, Michigan), containing various concentrations of PPT or bialaphos, ranging from 0 to 600 mg/L, to test the sensitivity of the pathogens. Four days after incubation, the radial lengths (mm) of the colonies were recorded to determine the sensitivity of the fungus to PPT or bialaphos. Linear regression was used to fit a line to the points and determine the concentration causing a 50% reduction in growth (ED_{50}). Statistical analysis of differences in the mean values of radial length of the fungi was achieved by Tukey's Honestly Significant Difference Test ($P = 0.05$).

Analysis of Bialaphos on Prevention of Pathogen Infection in Greenhouse

The uniformly grown transgenic plants and nontransgenic plants were inoculated with about 500 mg of autoclaved wheat seeds with actively growing mycelium of the fungi *Rhizoctonia solani*, *Sclerotinia homeocarpa*, or *Pythium aphanidermatum*, by evenly distributing on top of each plant. Various concentrations of bialaphos solution, ranging from 200 to 2400 mg/L, were sprayed on the creeping bentgrass either three hours before or two days after the fungus inoculation. After the bialaphos application, plants were wrapped with plastic bags with holes to raise the humidity. Transgenic and nontransgenic plants that were not treated with bialaphos were subject to the same pathogen infection procedures as well. Plant damage was visually rated on a 0 to 10 scale (0 = no damage; 5 = 50% plant damage; 10 = death) one week after inoculation. Data was also analyzed using Tukey's Honestly Significant Difference Test and F test (P = 0.05).

RESULTS AND DISCUSSION

Production of Transgenic Plants

A total of 38 independent, bialaphos- and PPT-resistant callus lines were recovered from 250 plates of five experiments after about three months of selection of the bombarded callus. The plantlets were regenerated from all callus lines one month after transfer to MS medium containing the same level of bialaphos or PPT.

Digested and undigested genomic DNA from the leaf samples of independently transgenic plants were subjected to Southern blot hybridization. The results revealed the presence of the *bar* gene in different transformation events. The results also indicated that the *bar* gene was integrated into genomic DNA with different integration sites and copy numbers in different transformation events. The identical patterns of hybridization bands in the six plants from the same transformation event confirmed that they were from the same transformed cell.

A 600 bp PAT-specific mRNA band was detected by Northern blot hybridization in analysis of total RNA from the young leaves of independently transformed plants, but not in untransformed control plants, confirming the accumulation of the *bar* transcripts in transgenic plants. The thin layer chromatography analysis of total protein extracts from young leaves of independent transgenic plants showed the PAT activity, indicating the expression of the *bar* gene.

The above results demonstrated that both bialaphos and PPT can be used as an efficient selective agent for creeping bentgrass transformation.

Herbicide Resistance and Accumulation of Ammonium

Plants regenerated from all independent transformation events were subjected to the spray of 1.2% Ignite®, the level that killed all of the control plants within 10 days. Except plants from one of the PPT-resistant callus lines, plants from the other 37 lines showed no obvious damage, indicating that the level of PAT activity in transgenic plants was sufficient enough to be resistant to at least 1.2% of Ignite®.

The results from evaluation of herbicide resistance levels of three independent transgenic events showed that the levels of PAT activity were different. LD_{50} values (lethal doses that caused 50% of plant death) of the three tested transgenic lines were about 20, 30, or 40 times higher than that of nontransgenic plants. The accumulation of ammonium was measured in the nontransgenic plants and the three tested transgenic lines after treatment with 1.2% Ignite®. The level of ammonium in control plants was normally between 14 and 20 µg NH_4^+-N per gram fresh sample. Ammonia accumulated up to more than 30-fold 2 hr after the application of Ignite® and reached 4,380 µg NH_4^+-N in control plants, which caused substantial necrosis in the plants. In the three tested transgenic lines, the levels of ammonium increased 10, 11, and 15 times, respectively, 2 hr after treatment. Although the levels of ammonium were able to decrease to normal two days after treatment, three transgenic lines showed different abilities of detoxification. The amount of ammonium in one transgenic line continued to accumulate for 6 hr and decreased to 12 hr after the Ignite® application, while plants from the other two lines started to decrease the accumulation of ammonium 4 hr after the application of Ignite®. The ability of these transgenic lines to detoxify the glufosinate-ammonium corresponded to their magnitude of herbicide resistance (LD_{50} values).

In conclusion, the transgenic plants showed a high level of resistance to PPT, which is the active ingredient of many commercial herbicides. The application of PPT- or bialaphos-containing herbicides on golf courses and fairways with transgenic creeping bentgrass will be able to provide a more efficient control of monocot and dicot weeds.

In Vitro Sensitivity of Pathogens to PPT and Bialaphos

The responses of fungal pathogens *Rhizoctonia solani* (brown patch), *Sclerotinia homeocarpa* (dollar spot), and *Pythium aphanidermatum* (*Pythium* blight) and creeping bentgrass to bialaphos or PPT were very different *in vitro*.

R. solani was very sensitive to bialaphos. Its radial mycelial growth was significantly suppressed by 1 mg/L of bialaphos. The fungal growth was reduced to 50% at about 5 mg/L of bialaphos (ED_{50} = 5.54 mg/L). A concentration of 60 mg/L bialaphos was able to inhibit the growth of *R. solani*. PPT was not as effective as bialaphos in suppressing the growth of *R. solani*, although a concentration of 25 mg/L PPT significantly inhibited the growth of mycelium.

To inhibit 50% of radial growth of *R. solani* mycelium, over 292 mg/L of PPT was required.

The response of mycelial growth of *S. homeocarpa* was also sensitive to the presence of bialaphos and PPT. The ED_{50} value of *S. homeocarpa* for bialaphos was 33.4 mg/L, higher than that of *R. solani*. The complete suppression of the mycelial growth of *S. homeocarpa* needed more than 200 mg/L of bialaphos. Unlike *R. solani, S. homeocarpa* was more sensitive to higher concentrations of PPT (ED_{50} = 270.06 mg/L). The effect of bialaphos or PPT amendment into PDA medium on the inhibition of mycelial growth of *R. solani* and *S. homeocarpa* was effective, with the highest concentration resulting in the least growth of mycelium.

P. aphanidermatum was the least sensitive fungus to both bialaphos and PPT as compared to *R. solani* and *S. homeocarpa.* Although 500 mg/L of bialaphos significantly reduced the mycelia (ED_{50} = 1467.18 mg/L), the whole plate was covered with the mycelium of *P. aphanidermatum* one week after inoculation. Up to 600 mg/L of PPT did not inhibit the radial mycelial growth of *P. aphanidermatum*. However, both bialaphos and PPT did inhibit the amount of mycelial growth of *P. aphanidermatum.*

In general, both bialaphos and PPT inhibited the radial and/or mycelial growth of *R. solani*, *S. homeocarpa*, and *P. aphanidermatum in vitro*. The inhibitory activity of bialaphos was superior to PPT. We speculated that the significant differences in the magnitude of ED_{50} values between bialaphos and PPT was due to the D-isomer of PPT, which might have interfered with the L-isomer of PPT in the binding of glutamine synthetase and reduced the inhibition efficiency of L-PPT.

Responses of Pathogens to Bialaphos on Greenhouse-Grown Transgenic Plants

After the application of various concentrations of bialaphos on transgenic creeping bentgrass, the effect of the three different pathogens on plant damages was analyzed. The application of bialaphos, even at one-tenth of the recommended herbicide rate (200 mg/L bialaphos) to kill common weeds, significantly suppressed the development of brown patch, 3 hr before or 2 days after inoculation of *R. solani*. The transgenic plants showed minimal damage, while control transgenic plants without spraying of bialaphos showed typical symptoms of brown patch disease and died 3 weeks after the pathogen inoculation. The difference in plant damage between two different timings of bialaphos application was not significant ($F = 0.29 < F_{0.05\,(1,\,63)} = 4.00$).

The *S. homeocarpa*-caused plant damage on transgenic bialaphos-resistant creeping bentgrass after the bialaphos application in two different applications were significantly less than those on transgenic plants not treated with bialaphos. However, the damage between the two application times was significant ($F = 8.23 > F_{0.05\,(1,\,81)} = 3.98$). When the bialaphos was sprayed on transgenic plants two days after the pathogen inoculation, there was more plant damage caused by the infection of *S. homeocarpa*. A 200 mg/L dose of bialaphos was also high enough to suppress the development of *S. homeocarpa*. Unlike the prevention of *R. solani* infection, the application of bialaphos was less effective to *S. homeocarpa,* and more plant damage was observed. The plant damage caused by brown patch or dollar spot was reduced by increasing the concentration of bialaphos applied. The disease development was significantly restrained and most plants were able to completely recover from the infection and grew normally.

Although 200 mg/L of bialaphos, applied 3 hr before inoculation of *Pythium* blight, significantly restrained the infection and reduced the plant damage, at least 800 mg/L of bialaphos was needed to have the same level of prevention on transgenic plants if bialaphos was applied 2 days after inoculation. Better control of *P. aphanidermatum* was carried out by using a high concentration of bialaphos and/or by applying bialaphos on transgenic plants before the pathogen inoculation ($F = 25.57 > F_{0.05\,(1,\,153)} = 3.96$).

The results demonstrated that bialaphos application can significantly suppress the development of brown patch, dollar spot, and *Pythium* blight. The application timing was important in better control of dollar spot and, particularly, in control of *Pythium* blight. Therefore, bialaphos can be used for simultaneous control of fungal infection and weed infestations in the fields of transgenic, bialaphos-resistant creeping bentgrass.

REFERENCES

Beard, J.B. 1982. *Turf management for golf courses.* Macmillan Publishing Company, NY. pp. 119–124.

Bayer, E., K.H. Gugel, K. Haebele, H. Hagenmaier, S. Jessipow, W.A. Koenig, and H. Zaehner. 1972. Phosphinothricin und phosphinothricyl-alanin. *Helv. Chim. Acta.*, 55:224–239.

DeBlock, M., J. Botterman, M. Vandewiele, J. Dockx, C. Thoen, V. Gosselé, N. R. Movva, C. Thompson, M. Van Montagu, and J. Leemans. 1987. Engineering herbicide resistance in plants by expression of a detoxifying enzyme. *EMBO J.*, 6:2513–2518.

D'Halluin, K., M. De Block, J. Denecke, J. Janssens, J. Leemans, A. Reynaerts, and J. Botterman. 1992. The *bar* gene as selectable and screenable marker in plant engineering. In *Methods in Enzymology*, 216, pp: 415–416. Wu, R., Ed., Academic Press, Inc., CA.

Hoerlein, G. 1994. Glufosinate (phosphinothricin), a natural amino acid with unexpected herbicidal properties. In: *Reviews of Environmental Contamination and Toxicology*, 138, pp. 73–145. Ware, G.W., Ed., Springer-Verlag, Inc., NY.

Kondo, Y., T. Shomura, Y. Ogawa, T. Tsuruoka, H. Watanabe, K. Totsukawa, T. Suzuki, C. Moriya, and J. Yoshida. 1973. Isolation and physico-chemical and biological characterization of SF-1293 substance. *Sci. Reports of Meiji Seika Kaisha.* 13:34–41.

Leason, M., D. Dunliffe, D. Parkin, P.J. Lea, and B.J. Miflin. 1982. Inhibition of pea leaf glutamine synthetase by methionine sulphoximine, phosphinothricin and other glutamate analogues. *Phytochem.,* 21:855–857.

Murakami, T., H. Anzai, S. Imai, A. Satoh, K. Nagaoka, and C.J. Thompson. 1986. The bialaphos biosynthetic genes of *Streptomyces hygroscopicus*: molecular cloning and characterization of the gene cluster. *Mol. Gen. Genet.,* 205:42–50.

Smiley, R.W. 1983. *Compendium of Turfgrass Diseases.* The American Phytopathological Society, St. Paul, MN, p. 11–72.

Southern, E.M. 1975. Detection of specific sequences among DNA fragments separated by gel electrophoresis. *J. Mol. Biol.,* 98:503–517.

Tachibana, K., T. Watanabe, Y. Sekizawa, and T. Takematsu. 1986. Action mechanism of bialaphos. II. Accumulation of ammonia in plants treated with bialaphos. *J. Pest. Sci.,* 11:33–37.

Thompson, C.J., N.R. Movva, R. Tizard, R. Crameri, J.E. Davies, M. Lauwereys, and J. Botterman. 1987. Characterization of the herbicide-resistance gene *bar* from *Streptomyces hygroscopicus. EMBO J.,* 6:2519–2523.

Uchimiya, H., M. Iwata, C. Nojiri, P.K. Samarajeewa, S. Takamatsu, S. Ooba, H. Anzai, A.H. Christensen, P.H. Quail, and S. Toki. 1993. Bialaphos treatment of transgenic rice plants expressing a *bar* gene prevents infection by the sheath blight pathogen (*Rhizoctonia solani*). *Bio/Technology,* 11:835–836.

Wadsworth, G.J., M.G. Redinbaugh, and J.G. Scandalios. 1988. A procedure for the small-scale isolation of plant RNA suitable for RNA blot analysis. *Analyt. Biochem.,* 172:279–283.

Wohlleben, M., W. Arnold, I. Broer, D. Hillemann, E. Strauch, and A. Puhler. 1988. Nucleotide sequence of the phosphinothricin *N*-acetyltransferase gene from *Streptomyces viridochromogenes* Tu494 and its expression in *Nicotiana tabacum. Gene.* 70:25–37.

Zhong, H., C. Srinivasan, and M.B. Sticklen. 1991. Plant regeneration via somatic embryogenesis in creeping bentgrass (*Agrostis palustris* Huds.). *Plant Cell Rep.* 10:453–456.

Zhong, H., M.G. Bolyard, C. Srinivasan, and M.B. Sticklen. 1993. Transgenic plants of turfgrass (*Agrostis palustris* Huds.) from microprojectile bombardment of embryogenic callus. *Plant Cell Rep.* 13:1–6.

Zhong, H., B. Sun, D. Warkentin, S. Zhang, R. Wu, T. Wu, and M.B. Sticklen. 1996. The competence of maize shoot meristems for integrative transformation and inherited expression of transgenes. *Plant Physiol.* 110:1097–1107.

Chapter 17

In Vitro Selection in *Agrostis stolonifera* var. *palustris*: Heat Tolerance and *Rhizoctonia solani* Resistance

J.V. Krans, S.L. Park, M. Tomaso-Peterson, and D.S. Luthe

ABSTRACT

In vitro selection was used to identify, isolate, and recover creeping bentgrass (*Agrostis stolonifera* var. *palustris* Farwell) variants with heat tolerance or resistance to *Rhizoctonia solani*. Tissue culture protocol for manipulating this species was: (a) callus induced from mature caryopses cultured on Murashige and Skoog (MS) medium with 1.0 mg 2,4-dichlorophenoxyacetic acid (2,4-D)/L in light (100/Em^2/sec^1 supplied by cool white fluorescent lamps) or dark incubation at 25°C; (b) callus maintained on MS medium with 1 to 5 mg 2,4-D/L in dark incubation at 25°C; and (c) shoots and roots formed from callus on MS medium with 0.10 mg 2,4-D and 1.0 mg kinetin/L in light incubation at 25°C. Plantlets transferred to soil showed 95% survival.

Suspension culture of callus prior to plating increased callus growth two-fold, but drastically inhibited plantlet formation from 400 to < 10 plantlets per dish. Large callus aggregates (0.5 to 1.0 mm diameter) used for plating had greater callus growth, but a lower yield of plantlets than small aggregates (0.25 to 0.5 mm diameter). Plating density of aggregates at 1 mg callus (50/mm^2 and

3 mg callus 50/mm^2 was optimum for inducing plantlet formation from large and small aggregates, respectively.

Heat-tolerant variants of creeping bentgrass were identified and isolated by culturing callus at 37, 39, or 40°C for 10 to 14 days depending on temperature. Plantlets were regenerated from surviving callus and screened in hydroponic culture at 40°C air (shoot) and solution (root) temperature for 3 weeks. Approximately 1% of plants subjected to this screen survived and were transferred as 5-cm plugs to a Tifgreen bermudagrass (*Cynodon dactylon* L. Pers. X *C. transvaalensis* Burtt-Davy) sod, mowed at 6 mm, and evaluated for growth and vigor. After 5 years, variants that displayed good growth and vigor were planted in an isolated polycross nursery in Oregon. From this nursery, progeny were collected and evaluated for growth and vigor. Based on parent and progeny performance, 22 variants were identified as superior types.

Heat-shock response was studied in heat-tolerant (SB) and non-tolerant variants (NSB). SB and NSB synthesized heat-shock proteins (HSPs) of 97, 83, 70, 40, 25, and 18 kD. Analysis by two-dimensional gel electrophoresis revealed SB synthesized two or three additional members of the HSP27 family, which were small (25kD) and more basic than those synthesized by NSB. A chi-square test of independence of F_1 progeny of NSB X SB indicated that heat tolerance and the presence of the additional HSP25 polypeptides were linked traits.

A novel *in vitro* selection technique, the Host-Pathogen Interaction System (HPIS), was used to identify and isolate *R. solani*–resistant variants in creeping bentgrass. Evaluation of the HPIS using *R. solani* isolates of varying pathogenicity revealed that the HPIS can be used to predict *R. solani* virulence. Sixty variants have been identified, isolated, and recovered with potential *R. solani* resistance using the HPIS. Nine of these variants plus a nonselected control were screened for *R. solani* resistance by inoculation in a controlled environment. Four variants displayed significantly less disease symptomology than the control. Field plot evaluation and inoculation of all 60 variants are currently under way.

INTRODUCTION

In vitro plant selection is a tissue culture process used to identify, isolate, and recover variant plant types. Plant tissue culture is the aseptic culture of excised plant parts, callus, organogenetic tissue, embryogenetic tissue, somatic embryos, or plantlets on semi-solid or liquid media. The manipulation of plants in tissue culture is achieved by altering the media and/or culture environment.

The *in vitro* selection process can identify and isolate desirable variants by using discriminating selection pressure. Discriminating pressures may include pathotoxins, environmental stresses, or other adverse conditions that affect the health and vigor of plants. Pressures are applied to callus when cell differentiation is low and traits are most vulnerable for selection.

Creeping bentgrass (*Agrostis stolonifera* var. *palustris* Farwell) is an important golf turf species. *In vitro* selection has been used in creeping bentgrass as a means to recover novel genotypes. This chapter reviews some of the procedures and findings of *in vitro* cell selection in creeping bentgrass.

MATERIALS AND METHODS

All tissue culture studies reported herein were conducted according to procedures previously reported (Bayta-Blanche, 1984; Blanche, 1986; Krans et al., 1982; Tomaso-Peterson, 1989; Zhao, 1992). Unless otherwise stated, "Penncross" creeping bentgrass was used in all studies.

Callus Induction, Maintenance, and Plantlet Regeneration

Protocols for callus induction, maintenance, and plantlet formation in creeping bentgrass have been previously reported (Blanche et al., 1986; Krans et al., 1982). These experiments evaluated: (a) various concentrations and combinations of 2,4-D and kinetin; (b) dark or light incubation; (c) callus age and prior culture environment; and (d) aggregate size and density of plated callus on callus induction, maintenance, and plantlet formation.

In Vitro Selection for Heat Tolerance

The identification, isolation, and recovery of heat-tolerant variants of creeping bentgrass have been previously reported (Bayta-Blanche, 1984). Steps for the *in vitro* selection of heat-tolerant variants were: (a) plated callus incubated on maintenance media at 37, 39, or 40°C for 10 to 14 days depending on temperature; (b) surviving heat-treated callus transferred to regeneration media to promote plant formation; (c) regenerated plantlets screened for heat tolerance in hydroponic culture at 40°C solution (root) and air (shoot) temperatures for 3 weeks; and (d) plants surviving hydroponic screen were transferred to soil in pots for field testing.

Parent and Progeny Evaluations for Heat Tolerance

Surviving plants from hydroponic heat screen tests were transferred as 5-cm-diameter plugs spaced on 60-cm centers into a Tifgreen bermudagrass (*Cynodon dactylon* L. Pers. X *C. transvaalensis* Burtt-Davy) turf mowed at 6-mm cutting height three times weekly. Field plot evaluation was in either of two germplasm nurseries, depending upon date of recovery. Nursery I was planted in March, 1987 and Nursery II was planted in April 1989. The experimental design was a randomized complete block with three replications. Plug growth was measured after frost each year (mid-November) when bermudagrass was dormant and creeping bentgrass was highly visible. During the evaluation period, plots received minimal irrigation, no pesticide treatments, and 5 g N/m/yr. Plant performance was determined by monitoring the size of the bentgrass plugs annually over a 5-year period.

In 1989, 48 variants were selected from Nursery I and planted in an isolated polycross nursery in Oregon. In 1990, 5 variants were selected from Nursery II and planted in a second isolated polycross nursery in Oregon. Polycross progeny of each of the 48 seed parents from Nursery I were planted in October 1990. In October 1991, polycross progeny of the 5 seed parents from Nursery II were planted. Both progeny trials were planted in Mississippi in 60/cm^2 plots replicated three times, and plots were maintained under putting green culture. Turf performance rating (visual score of 1 to 9; with 1 = poor and 9 = excellent) of Nursery I progeny was taken August 31, 1991 and Nursery II progeny was taken September 15, 1992. In the first polycross progeny trial (planted 1990), Penncross was planted as a control, and in the second trial (planted 1991), Penncross and Southshore were planted as controls.

Heat Shock Response

Heat shock response of an *in vitro* selected heat-tolerant (SB) and non-tolerant (NSB) variant of creeping bentgrass and their progeny has been reported (Park et al., 1996). SB and NSB variants were selected among a population of variants generated by Bayta-Blanche (1984) and evaluated by Kemp (1987). Reciprocal crosses were made between SB and NSB by Ms. Crystal Fricker, Turfgrass Breeder, Pure Seed Testing, Inc., Hubbard, Oregon. The SB, NSB, and progeny were increased vegetatively as needed. SB and NSB were evaluated for heat tolerance in a 40°C hydroponic heat screen according to Bayta-Blanche (1984). Qualitative analyses of heat shock proteins (HSPs) were conducted using leaf segments radiolabeled by the addition of ^{35}S-methionine and ^{35}S-cysteine (trans ^{35}S label reagent). Samples were analyzed by 1-D (one-dimensional) and/or 2-D (two-dimensional) sodium dodecylsulfate (SDS) polyacrylamide gel electrophoresis (PAGE). The temperature and time required for HSP induction of SB and NSB were determined by incubation and equilibration of leaf segments at temperatures ranging from 25 to 40°C, then labeled for analysis.

In Vitro Selection for *R. solani* Resistance

The Host-Pathogen Interaction System (HPIS) has been used for the *in vitro* selection of *R. solani*–resistant variants of creeping bentgrass. The description, use, and application of the HPIS have been reported (Tomaso-

Peterson, 1989; Tomaso-Peterson and Krans, 1990). The HPIS was evaluated using Penncross callus and 13 *R. solani* isolates of known virulence provided by Phil Cobaugh, Plant Pathologist, Texas A&M University, Dallas. Callus mortality was determined using a triphenyltetrazolium chloride stain. Callus mortality and isolate virulence ratings were evaluated using logistic regression analysis to model the relationship between callus mortality and the probability of virulence on whole plants.

Potentially resistant variants were recovered using the HPIS and a highly virulent *R. solani* isolate (RVPI) provided by Dr. H.B. Couch, Virginia Polytechnic Institute and State University, Blacksburg. Steps used in the recovery of these potentially resistant plants were: (a) callus/RVPI co-culture in the HPIS for 24 hr; (b) treated callus transferred to regeneration medium; (c) regenerated plantlets/RVPI co-cultured in HPIS plantlet chamber; and (d) surviving plantlets transferred to soil in pots for planting in the field. Nine of 60 surviving plants were screened for resistance under controlled conditions by inoculating with *R. solani* hyphal plugs. After 41 hr of incubation at 30°C and high humidity, plants were transferred to 20°C and low humidity to halt disease progress. Disease severity among variants was determined by measuring diameter of necrotic grass caused by infection.

In 1995, over 60 variants recovered from HPIS were planted in 60/cm^2 field plots replicated three times and maintained under putting green culture. Plots are scheduled for field inoculation with *R. solani* in July 1996.

RESULTS AND DISCUSSION

Callus Induction, Maintenance, and Plantlet Regeneration

Penncross creeping bentgrass has been effectively manipulated in tissue culture (Krans et al., 1982). Murashige and Skoog (MS) medium with 1.0 to 10.0 mg 2,4-dichlorophenoxyacetic acid (2,4-D)/L in light (100/Em^2/sec supplied by cool white fluorescent lamps) or dark incubation promoted callus induction from caryopses. Callus was best maintained on MS medium with 1 to 5 mg 2,4-D/L in dark incubation. Shoot and root formation was promoted from callus on MS medium with 0.10 mg 2,4-D and 1.0 mg kinetin/L in light incubation. Plantlets potted to soil had 95% survival.

Caryopses incubated in the dark yielded more callus than those incubated in the light. Although kinetin supplements were not required for callus induction, callus formation was greater in the dark on medium with 0.01 mg kinetin/L than medium without kinetin when combined with 2,4-D. Callus was initiated from the scutellum, the root cortex region, and along the coleoptile up to and adjacent to the first node. The greater callus formation in dark than light incubation was attributed to the elongation of the first internode in the dark and corresponding increased surface area for callus induction.

Plantlet formation from callus was predisposed by prior culture conditions (Blanche et al., 1986). Plantlet formation was inhibited when callus was previously cultured with 5 mg or more 2,4-D/L. Culturing callus in suspension prior to plating increased callus growth twofold, but drastically inhibited plantlet formation from 400 to < 10 plantlets/dish (Table 17.1). Large callus aggregates (0.5 to 1.0 mm diameter) used for plating had greater callus growth, but a lower yield of plantlets than small aggregates (0.25 to 0.5 mm diameter) (Table 17.2). Plating density of aggregates at 1 mg and 3 mg of callus 50/mm^2 was optimum for inducing plantlets. Hormone requirements for optimum plantlet formation from callus were 0.10 mg 2,4-D and 1.0 mg kinetin/L.

In Vitro Selection for Heat Tolerance

Plated callus subjected to high temperature treatments of 37, 39, and 40°C for 2-week durations over a 2-year period generated approximately 2500 potential heat-tolerant variants. This variant population subjected to a hydroponic heat screen resulted in 240 surviving plants. Of this population, 211 were planted in Nursery I and 29 were planted in Nursery II.

Parent and Progeny Evaluations for Heat Tolerance

After 5 years, 50 of 211 variants planted in Nursery I and 9 of 29 variants planted in Nursery II encroached into the bermudagrass turf (plug diameter > 6 cm).

Progeny evaluation of 48 variants from Nursery I revealed that progeny of 23 variants rated better in turf performance (turf scores 5.3 or greater) than Penncross (turf score 2.7, LSD 0.05 = 2.9). In a second progeny evaluation trial using parents from Nursery II, progeny of 2 variants rated better in turf performance (turf scores 4.7 or greater) than Penncross (turf score 3.0, LSD 0.05 = 1.7). Based on parent and progeny turf performance, 22 parents have been identified as superior types with potential for cultivar development.

Heat Shock Response

In creeping bentgrass, analysis for HSP in SB and NSB using 1-D gel electrophoresis at heat shock (40°C) or control (23°C) temperatures revealed the presence of two groups of HSPs, a high-molecular-mass HSP group (60–110 kD) and a low-molecular-mass HSP group (16–32 kD). There were no major differences in the time or temperature required to induce HSP synthesis in the two variants.

Analysis for HSPs by 2-D gel electrophoresis revealed two qualitative differences between NSB and SB. In NSB, two additional members of the

Table 17.1. Callus Growth and Plantlet Formation from Creeping Bentgrass Callus not Previously Cultured in Suspension and Callus Cultured in Suspension for 12 and 24 Days (Blanche et al., 1986).

Callus Source	Media Type[a]	Culture Period[b]	Culture Time[c] (Weeks)	Light or Dark Culture	Callus Growth[d] (Fold Increase)	Plantlet Formation[e] (No.)
SC	MM	First	6	Dark	9.3	0
	RM	Second	4	Light	NR[f]	293
SC	RM	First	5	Light	NR	400
	RM	Second	5	Light	NR	410
SC in suspension for 12 days	RM	First	5	Light	1.2	0
	RM	Second	5	Light	NR	9
SC in suspension for 24 days	MM	First	6	Dark	18.1	0
	RM	Second	4	Light	NR	3
SC in suspension for 24 days	RM	First	5	Light	1.3	0
	RM	Second	5	Light	NR	6
LSD	(0.05)				0.4	20

[a] Maintenance media (MM) consisted of solid media with 5 mg 2,4-D/L; and regeneration media (RM) consisted of solid media with 0.01 mg 2,4-D/L and 0.10 mg kinetin/L.
[b] First culture period refers to the incubation of callus after initial planting; second culture period refers to the incubation of callus after transfer from the first culture period.
[c] Culture time refers to the length of time callus was incubated on either media type.
[d] Callus growth based on callus wet weights.
[e] Plantlet formation based on shoot number per plate > 2 mm.
[f] NR refers to callus wet weights not recorded due to plantlets forming in the cultures.

Table 17.2. Callus Growth and Plantlet Formation from Stock Callus (SC) Plated at Two Aggregate Sizes and Three Plating Densities (Blanche et al., 1986).

Aggregate Size (mm)	Plating Densities (mg•50 mm)	Callus Growth[a] (Fold Increase)	Plantlet Formation[b] (No.)
0.50 to 1.0	1	21.6	390
	3	14.2	120
	5	10.6	70
0.25 to 0.50	1	13.9	330
	3	11.5	390
	5	9.9	260
	LSD (0.05)	1.3	129

[a] Callus growth based on callus wet weights after culturing for 6 weeks in the dark with 5 mg 2,4-D/L.
[b] Plantlet formation based on shoot number per plate > 2 mm.

HSP18 family were synthesized. In SB, two or three additional members of the HSP27 family, which were smaller (HSP25 kD) and more basic, were synthesized.

Reciprocal crosses between SB and NSB generated 20 progeny (19 from NSB X SB and 1 from SB X NSB). Based on 2-D gel electrophoresis, seven of the 19 F_1 progeny of NSB X SB did not synthesize the HSP25 polypeptides (Table 17.3). The single progeny from SB X NSB did synthesize the HSP polypeptide, which indicates that the trait was not maternally inherited. Thermotolerance of F_1 progeny determined by hydroponic screening showed that the F_1 progeny that synthesized the HSP25 polypeptides were significantly ($P < 0.01$) more heat-tolerant than progeny that did not synthesize the HSP25 polypeptides (Table 17.4). The presence of HSP25 polypeptides in F_1 progeny was positively correlated ($r = 0.743$, $P < 0.01$) with thermotolerance. A Chi-square test of independence between the presence of the HSP25 polypeptides and the level of heat tolerance showed a significant relationship ($\chi^2 = 22.45$, $P < 0.01$) (Table 17.5). There was no apparent relationship between the synthesis of the additional HSP18 polypeptides in NSB and thermotolerance. These findings indicate that the presence of the additional HSP25 polypeptides and heat tolerance are linked.

In Vitro Selection for *R. solani* Resistance

An evaluation of the HPIS revealed that this system can predict *R. solani* infection in creeping bentgrass. Based on a logistic regression analysis model

Table 17.3. Leaf Damage Scores[a] of F_1 Progeny of NSB X SB (Nos. 1–19) and SB X NSB (no. 20) After 9 Days of Hydroponic Heat Stress at 40°C (Park et al., 1996).

F_1 Progeny	Extra HSP25	Damage Score[b]							
7	–	9.4	A						
16	–	9.0	A						
17	–	8.3	A						
11	–	7.2	A	B					
18	–	6.7	A	B					
19	–	5.4		B	C	D			
8	+	5.4		B	C	D	E		
14	+	5.4		B	C	D	E		
1	+	4.9		B	C	D	E	F	
6	+	4.9		B	C	D	E	F	G
3	+	4.8		B	C	D	E	F	G
2	+	4.5		B	C	D	E	F	G
5	+	4.4			C	D	E	F	G
13	+	4.2			C	D	E	F	G
10	–	4.2			C	D	E	F	G
9	+	3.3				D	E	F	G
12	+	3.3				D	E	F	G
15	+	2.7					E	F	G
4	+	2.4						F	G
20	+	2.1							G

[a] A damage score of 1 was given when less than one-half of the leaf blade contained necrotic regions. A damage score of 5 was given when more than one-half of the leaf blade contained necrotic regions, but green areas were still apparent. A damage score of 10 was given when the leaf blade was totally necrotic and desiccated. The scores for each leaf blade of the plant were added and divided by the number of leaf blades.

[b] Means with the same letter are not significantly different (LSD=2.74). Statistical analysis was done using Fisher's protected LSD.

of the relationship between callus mortality and pathogen virulence, the probability of the HPIS to predict virulence of *R. solani* isolates was significant ($p = 0.03$) (Figure 17.1). Sixty potentially *R. solani*–resistant variants were recovered using the HPIS. Nine of these were screened for *R. solani* resistance and 4 displayed enhanced resistance (Table 17.6). All sixty variants are now planted in the field and currently undergoing *R. solani* field screening.

SUMMARY

In vitro selection in creeping bentgrass is a means of recovering new and novel plant types. Protocol for the tissue culture manipulation of Penncross

Table 17.4. Mean Heat Damage Scores for F_1 Progeny of NSB X SB and SB X NSB With and Without the Extra HSP25 Polypeptides (LSD=0.9848) (Park et al., 1996).

Extra HSP25	Mean Damage Score[a]
–	7.17
+	4.02

[a] A damage score of 1 was given when less than one-half of the leaf blade contained necrotic regions. A damage score of 5 was given when more than one-half of the leaf blade contained necrotic regions but green areas were still apparent. A damage score of 10 was given when the leaf blade was totally necrotic and desiccated. The scores for each leaf blade of the plant were added and divided by the number of leaf blades.

Table 17.5. Distribution of Damage Scores for 120 F_1 Progeny With or Without the Additional HSP25 Polypeptides. (χ^2=22.45 [$P < 0.01$]).

	No. of F_1 plants	
Heat Tolerance[a]	+HSP25	–HSP25
High damage	14	24
Intermediate damage	29	13
Low damage	35	5

[a] A damage score of 1 was given when less than one-half of the leaf blade contained necrotic regions. A damage score of 5 was given when more than one-half of the leaf blade contained necrotic regions but green areas were still apparent. A damage score of 10 was given when the leaf blade was totally necrotic and desiccated. The scores for each leaf blade of the plant were added and divided by the number of leaf blades. The higher the score, the greater the damage. High damage was a score of 7.5; intermediate damage was a score of $> 2.5 \leq 7.5$; low damage was a score of 2.5.

has been well defined. However, callus of Penncross can be easily predisposed to limited plantlet formation. Caution should be exercised in handling and maintaining Penncross callus to avoid inhibition of plantlet formation.

All variants generated from heat selection research are in repository for use in developing a synthetic cultivar. The elucidation of heat shock response in creeping bentgrass adds new information to the heat tolerance mechanisms in this species. Monitoring the presence or absence of the two or more additional members of the HSP27 family, linked to heat tolerance, may be useful in screening germplasm.

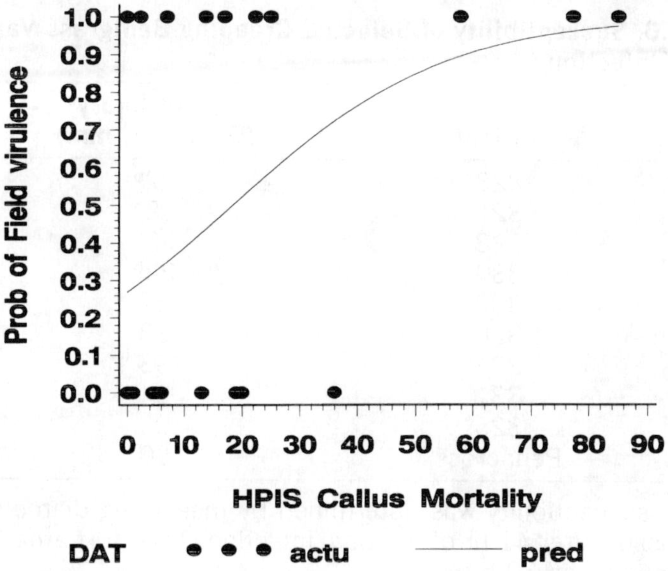

Figure 17.1. Predicted probability of *R. solani* virulence vs. HPIS callus mortality (0 = nonvirulent and 1 = virulent).

The HPIS has the potential to recover germplasm with enhanced *R. solani* resistance in creeping bentgrass. Field trials now under way will help to further determine the effectiveness of the HPIS.

REFERENCES

Bayta-Blanche, F.C.B. 1984. Evaluation of cell selection schemes with salt as a selection pressure in creeping bentgrass. MS thesis. Mississippi State University, Starkville.

Blanche, F.C., J.V. Krans, and G.E. Coats. 1986. Improvement in callus growth and plantlet formation in creeping bentgrass. *Crop Sci.,* 26:1245–1248.

Jakob, U., M. Gaestel, K. Engel, and J. Buchner. 1993. Small heat shock proteins are molecular chaperones. *J. Biol. Chem.,* 286:1517–1520.

Kemp, M.L. 1987. The field performance and breeding characteristics of selected creeping bentgrass variants. MS thesis. Mississippi State University, Starkville.

Krans, J.V., V.T. Henning, and K.C. Torres. 1982. Callus induction, maintenance, and plantlet regeneration in creeping bentgrass. *Crop Sci.,* 22:1193–1197.

Park, S.L., S. Renuka, J.V. Krans, and D.S. Luthe. 1996. Heat-shock response in heat-tolerant and nontolerant variants of *Agrostis palustris* Huds. *Plant Physiol.,* 111:515–524.

Table 17.6. Susceptibility of Selected Creeping Bentgrass Variants to *R. solani* Infection.

Variant	Size of Injury[a] (Diameter, mm)
323	22[b]
327	28[b]
328	25[b]
330	28[b]
332	32
326	33
324	33
331	34
325	35
Penncross	50

[a] Disease susceptibility was determined by measuring diameter of necrotic tissue after 41 hr of *R. solani* infection. Total turf area available for infection was 60 mm.

[b] Significantly (p=0.05) different from other values based on General Linear Model analysis.

Tomaso-Peterson, M. 1989. Evaluation of a new *in vitro* cell selection technique. MS thesis. Mississippi State University, Starkville.

Tomaso-Peterson, M. and J.V. Krans. 1990. Evaluation of a new *in vitro* cell selection technique. *Crop Sci.,* 30:226–229.

Vierling, E. The roles of heat shock proteins in plants. 1991. *Ann. Rev. Plant Physiol. Plant Mol. Biol.* 42:579–620.

Zhao, L. 1992. Refinement and evaluation of the Host-Pathogen Interaction System (HPIS) using *Rhizoctonia solani* and "Penncross" creeping bentgrass. MS thesis. Mississippi State University, Starkville.

Chapter 18

Biotechnology in Fescues and Ryegrasses: Methods and Perspectives

G. Spangenberg, Z.-Y. Wang, and I. Potrykus

INTRODUCTION

The two closely related genera *Festuca* L. (fescues) and *Lolium* L. (ryegrasses) include some well-adapted, persistent grass species that are widely used for forage, conservation, and turf purposes. Specifically, tall fescue (*F. arundinacea*), meadow fescue (*F. pratensis*), red fescue (*F. rubra*), Italian ryegrass (*L. multiflorum*) and perennial ryegrass (*L. perenne*) are of particular relevance, since they form the foundation of grassland agriculture in temperate climates throughout the world (Jauhar, 1993).

Genetic improvement of these allogamous wind-pollinated fescues and ryegrasses by conventional plant breeding is slow since self-infertility limits inbreeding to concentrate desired genes for use in rapid development of new cultivars. Therefore, biotechnological approaches such as genetic transformation for the direct introduction of useful genes, and molecular markers as tools in cultivar identification and marker-assisted selection, show promise when considered as part of fescue and ryegrass improvement programs.

Significant progress has been made in establishing the methodological basis required for the genetic manipulation of fescues and ryegrasses in recent years. Our contribution to the following areas: (a) plant regeneration from suspension cells and protoplasts, (b) symmetric and asymmetric somatic hybridization, and (c) gene transfer methods and recovery of transgenic plants, is outlined here for key species within the *Festuca-Lolium* complex.

PLANT REGENERATION FROM SUSPENSION CELLS AND PROTOPLASTS

An efficient system for regeneration of soil-grown plants from embryogenic suspension cells and corresponding protoplasts has been worked out for different *Festuca* and *Lolium* species: *F. arundinacea* (Wang et al., 1995), *F. pratensis* (Wang et al., 1993a), *F. rubra* (Spangenberg et al., 1994a), *L. multiflorum* (Wang et al., 1993b), *L. perenne* (Wang et al., 1993b, 1995), and *L. boucheanum* (Wang et al., 1993b). These reproducible and efficient plant regeneration systems are based on: (a) a genotype screening within different cultivars for the identification of embryogenic callus of single-seed origin; (b) the establishment of single genotype-derived highly embryogenic cell suspensions; (c) the cryopreservation of embryogenic suspension cultures for their long-term storage under liquid nitrogen; (d) the isolation of morphogenic protoplasts from young highly embryogenic cell suspensions; and (e) the culture of totipotent protoplasts in a bead-type system, including fast-growing non-morphogenic nurse cells. Overall protoplast plating efficiencies varied between 10^{-3} and 10^{-4}. Plant regeneration frequencies from suspension cells- and protoplasts-derived calli were in the range of 5% to 90%, depending on the species and cultivar.

A reproducible method for the cryopreservation of the single genotype-derived embryogenic cell suspensions was established for the different *Festuca* and *Lolium* species considered (Wang et al., 1994). The protocol allows for a long-term availability of highly regenerable embryogenic suspension cultures. Different parameters, such as cryoprotectant composition, pre-freezing osmotic adaptation of suspension cultures, cooling regimes, and post-thaw washing of cryopreserved embryogenic cultured cells, were partially optimized. Using this protocol, 40% to 70% of the cryopreserved and thawed cells were able to reinitiate post-thaw growth, to maintain their embryogenic character, and to regenerate into mature plants. Embryogenic cell suspensions were re-established from frozen-thawed cultures. Plants were regenerated from cryopreserved and re-established embryogenic cell suspensions and corresponding protoplasts at frequencies comparable to those of the original non-frozen cultures (Wang et al., 1994).

Plants recovered from maintained and/or cryopreserved and re-established suspension cultures and corresponding protoplasts in *F. arundinacea, F. rubra, F. pratensis, L. perenne* and *L. multiflorum* were established in soil, brought to flower and set seeds (Wang et al., 1993a,b, 1994, 1995; Spangenberg et al., 1994a). A RAPD analysis on the genetic stability of protoplast- and suspension cell–derived plants in *F. pratensis, F. arundinacea* and *L. multiflorum* revealed limited newly induced genetic variability at the loci screened with molecular markers (Vallés et al., 1993; Wang et al., 1993b, 1994).

On this experimental basis two types of genetic manipulations have been described: protoplast fusion, and genetic transformation mediated by direct gene transfer to protoplasts or by microprojectile bombardment of embryogenic suspension cells, for the recovery of somatic hybrid/cybrid and transgenic plants, respectively.

SYMMETRIC AND ASYMMETRIC SOMATIC HYBRIDIZATION

Genotypically and phenotypically different somatic *Festulolium* hybrid plants have been regenerated from symmetric and asymmetric protoplast fusions (Takamizo et al., 1991; Takamizo and Spangenberg, 1994; Spangenberg et al., 1994b, 1995a). Intergeneric symmetric and asymmetric somatic hybrids were obtained in the combinations *F. arundinacea* (+) *L. multiflorum* and *F. rubra* (+) *L. perenne*. Metabolically inactivated protoplasts isolated from embryogenic suspension cultures of *F. arundinacea* or *F. rubra* were electrofused with unirradiated (symmetric fusion) or X-ray irradiated (asymmetric fusion) protoplasts prepared from non-morphogenic cell suspensions of *L. multiflorum* or *L. perenne* (Spangenberg et al., 1995a; Legris et al., submitted). The true somatic hybrid nature of the plants recovered was demonstrated by chromosome counts, isozyme patterns, genomic Southern blots, quantitative dot blot and *in situ* hybridizations. For the analysis of the nuclear composition of these symmetric and asymmetric somatic hybrids, species-specific repetitive nuclear DNA sequences from *F. arundinacea*, *L. multiflorum,* and *F. rubra* were isolated and characterized, and found to be dispersed and evenly represented in the respective genomes (Pérez-Vicente et al., 1992; Spangenberg et al., 1995a).

The analysis of asymmetric somatic hybrids arising from "donor-recipient" protoplast fusions revealed that the X-ray irradiation treatment of the donor protoplasts favored the unidirectional elimination of most or few of the donor chromosomes in the somatic hybrid plants recovered. A bias toward recipient-type organelles was apparent in highly asymmetric somatic hybrids that showed an extensive (>80%) donor nuclear genome elimination induced by the X-ray irradiation (Spangenberg et al., 1994b, 1995a). These results demonstrated for two intergeneric combinations within the *Festuca-Lolium* complex that asymmetric protoplast fusion allows for directed one-step partial nuclear genome transfer. These procedures show potential for facilitating limited alien gene transfer between sexually (in)compatible species to complement or enhance conventional wide hybridization programs in forage- and turf-type fescues and ryegrasses.

GENE TRANSFER METHODS AND REGENERATION OF TRANSGENIC PLANTS

Transgenic plants have been obtained by direct gene transfer to protoplasts with overall frequencies between 10^{-4} and 10^{-6} in *F. arundinacea* (Wang et al., 1992), *F. rubra* (Spangenberg et al., 1994a), and *F. pratensis* (Spangenberg et al., 1995b). Chimeric hygromycin phosphotransferase (HPH) gene (*hph*) driven by enhanced CaMV 35S promoter and rice *Act1* 5' regulatory sequences were used as selectable markers. Regenerated plants were screened by PCR and grown until maturity under containment glasshouse conditions. Their transgenic nature was demonstrated by Southern hybridization analysis, HPH enzyme assays and *in situ* hybridization to metaphase chromosomes (Wang et al., 1992; Spangenberg et al., 1994a, 1995b).

Analogous experiments on direct gene transfer to protoplasts of *L. multiflorum* and *L. perenne* have not led to the recovery of transgenic plants.

Transgenic plants have been obtained by microprojectile bombardment of embryogenic suspension cells in *F. arundinacea* (Spangenberg et al., 1995c), *F. rubra* (Spangenberg et al., 1995c), *L. perenne* (Spangenberg et al., 1995d) and *L. multiflorum* (Ye et al., 1997). Chimeric hygromycin phosphotransferase gene constructs driven by CaMV 35S promoter, rice *Act1,* or maize *Ubi1* 5' regulatory sequences were used. Bombardment parameters of embryogenic suspension cells with the cost-effective particle inflow gun were partially optimized using transient expression assays of a chimeric β-glucuronidase (GUS) gene (*gusA*) construct. The DNA-particle delivery parameters that reproducibly allowed the highest number of transient GUS expression events in bombarded suspension cells were: 500 μm baffle mesh size; 12 cm baffle, and 15 cm target-bombardment distances; 6 bar bombardment pressure; one shot with a 10 μL DNA-spermidine bound particle suspension per target using 0.5 mg gold particles coated with 10 μg plasmid DNA. For the recovery of stably transformed clones, hygromycin selection using liquid and solidified media was tested. Initial selection in liquid culture medium allowed for a twofold — compared with continuous plate selection using solid medium — recovery efficiency of transformed hygromycin-resistant clones. For the fescues, on average, three hygromycin-resistant calli were recovered in 39% and 45% of the bombarded dishes in *F. arundinacea* (136 calli/117 dishes) and *F. rubra* (131 calli/97 dishes), respectively (Spangenberg et al., 1995c). For the ryegrasses, one resistant callus was obtained in 38% and 59% of the bombarded dishes in *L. perenne* (36 calli/96 dishes) and *L. multiflorum* (93 calli/156 dishes), respectively (Spangenberg et al., 1995d; Ye et al., 1997). Plants were regenerated from 35% and 85% of the hygromycin-resistant calli obtained in tall and red fescue, respectively. Regeneration frequencies from hygromycin-resistant calli were 23% for perennial ryegrass and 33% in Italian ryegrass. *In vitro*–regenerated plantlets were screened by PCR. The transgenic nature of the regenerated plants was confirmed by Southern hybridization analysis, thus indicating that the selection schemes used were tight. Expression of the transgene in transformed mature plants was demonstrated by Northern analysis and HPH enzyme assays.

Furthermore, the required experimental basis for the delivery of microprojectiles coated with DNA to L1 and L2 cells of floral and vegetative meristems has been established for two *Lolium* species (Pérez-Vicente et al., 1993).

PROSPECTS

As outlined, methods required for the genetic manipulation of key forage and turfgrasses within the *Festuca-Lolium* complex are now in place. They have opened up opportunities for evaluating the potential of different experimental strategies directed to achieve genetic engineering objectives defined in a sensible manner for specific production systems. These will presumably include, first, particular nutritional improvements (e.g., improved digestibility by reduced/altered lignin, increased non-structural carbohydrate content, regulated expression of "rumen bypass" proteins rich in essential amino acids), pest and pathogen protection (e.g., regulated expression of *Bt* insecticidal proteins, proteinase inhibitors and antifungal proteins, coat protein–mediated virus resistance, surrogate transformation with genetically engineered "ruminant-safe" endophytes), and aspects of growth and development (e.g., delayed flowering and senescence, low-allergen pollen, cytoplasmic male sterility and apomixis).

REFERENCES

Jauhar, P.P. 1993. Cytogenetics of the *Festuca-Lolium* complex: Relevance to breeding. In *Monographs on Theoretical and Applied Genetics,* Vol. 18. Frankel, R., M. Grossman, H.F. Linskens, P. Maliga, and R. Riley, Eds., Springer, Berlin.

Legris, G., Z.Y. Wang, I. Potrykus, and G. Spangenberg. Asymmetric somatic hybridization between red fescue (*Festuca rubra* L.) and irradiated perennial ryegrass (*Lolium perenne* L.) protoplasts. *Submitted.*

Pérez-Vicente, R., L. Petris, M. Osusky, I. Potrykus, and G. Spangenberg. 1992. Molecular and cytogenetic characterization of repetitive DNA sequences from *Lolium* and *Festuca*: Applications in the analysis of *Festulolium* hybrids. *Theor. Appl. Genet.* 84, pp. 145–154.

Pérez-Vicente, R., X.D. Wen, Z.Y. Wang, N. Leduc, C. Sautter, E. Wehrli, I. Potrykus, and G. Spangenberg. 1993. Culture of vegetative and floral meristems in ryegrasses: potential targets for microballistic transformation. *J. Plant Physiol.* 142:610–617.

Spangenberg, G., Z.Y. Wang, J. Nagel, and I. Potrykus. 1994a. Protoplast culture and generation of transgenic plants in red fescue (*Festuca rubra* L.). *Plant Sci.* 97:83–94.

Spangenberg, G., M.P. Vallés, Z.Y. Wang, P. Montavon, J. Nagel, and I. Potrykus. 1994b. Asymmetric somatic hybridization between tall fescue (*Festuca arundinacea* Schreb.) and irradiated Italian ryegrass (*Lolium multiflorum* Lam.) protoplasts. *Theor. Appl. Genet.* 88:509–519.

Spangenberg, G., Z.Y. Wang, G. Legris, P. Montavon, T. Takamizo, R. Pérez-Vicente, M.P. Vallés, J. Nagel, and I. Potrykus. 1995a. Intergeneric symmetric and asymmetric somatic hybridization in *Festuca* and *Lolium*. *Euphytica.* 85:235–245.

Spangenberg, G., Z.Y. Wang, M.P. Vallés, and I. Potrykus, 1995b. Transformation in tall fescue (*Festuca arundinacea* Schreb.) and meadow fescue (*Festuca pratensis* Huds.). In *Biotechnology in Agriculture and Forestry*, Vol. 34, Bajaj, Y.P.S., Ed., Springer, Berlin.

Spangenberg, G., Z.Y. Wang, X.L. Wu, J. Nagel, V.A. Iglesias, and I. Potrykus. 1995c. Transgenic tall fescue (*Festuca arundinacea*) and red fescue (*F. rubra*) plants from microprojectile bombardment of embryogenic suspension cells. *J. Plant Physiol.* 145:693–701.

Spangenberg, G., Z.Y. Wang, X.L. Wu, J. Nagel, and I. Potrykus. 1995d. Transgenic perennial ryegrass (*Lolium perenne*) plants from microprojectile bombardment of embryogenic suspension cells. *Plant Sci.* 108:209–217.

Takamizo, T. and G. Spangenberg. 1994. Somatic hybridization in *Festuca* and *Lolium*. In *Biotechnology in Agriculture and Forestry*, Vol. 27. Bajaj, Y.P.S., Ed., Springer, Berlin.

Takamizo, T., G. Spangenberg, K. Suginobu, and I. Potrykus. 1991. Intergeneric somatic hybridization in Gramineae: Somatic hybrid plants between tall fescue (*Festuca arundinacea* Schreb.) and Italian ryegrass (*Lolium multiflorum* Lam.). *Mol. Gen. Genet.* 231:1–6.

Vallés, M.P., Z.Y. Wang, P. Montavon, I. Potrykus, and G. Spangenberg. 1993. Analysis of genetic stability of plants regenerated from suspension cultures and protoplasts of meadow fescue (*Festuca pratensis* Huds.). *Plant Cell Rep.* 12:101–106.

Wang, Z.Y., T. Takamizo, V.A. Iglesias, M. Osusky, J. Nagel, I. Potrykus, and G. Spangenberg. 1992. Transgenic plants of tall fescue (*Festuca arundinacea* Schreb.) obtained by direct gene transfer to protoplasts. *Bio/Technology.* 10:691–696.

Wang, Z.Y., M.P. Vallés, P. Montavon, I. Potrykus, and G. Spangenberg. 1993a. Fertile plant regeneration from protoplasts of meadow fescue (*Festuca pratensis* Huds.). *Plant Cell Rep.* 12:95–100.

Wang, Z.Y., J. Nagel, I. Potrykus, and G. Spangenberg. 1993b. Plants from cell suspension-derived protoplasts in *Lolium* species. *Plant Sci.* 94:179–193.

Wang, Z.Y., G. Legris, J. Nagel, I. Potrykus, and G. Spangenberg. 1994. Cryopreservation of embryogenic cell suspensions in *Festuca* and *Lolium* species. *Plant Sci.* 103:93–106.

Wang, Z.Y., G. Legris, M.P. Vallés, I. Potrykus, and G. Spangenberg. 1995. Plant regeneration from suspension and protoplast cultures in the temperate grasses *Festuca* and *Lolium*. In *Current Issues in Plant Molecular and Cellular Biology*. Terzi, M., R. Cella, and A. Falavigna, Eds., Kluwer Academic Publishers, Dordrecht, The Netherlands.

Ye, X.D., Z.Y. Wang, X.L. Wu, I. Potrykus, and G. Spangenberg. Transgenic Italian ryegrass (*Lolium multiflorum*) plants from microprojectile bombardment of embryogenic suspension cells. *Plant Cell Rep.* 16:379–384.

Chapter 19

In Vitro Culture, Somaclonal Variation, and Transformation Strategies with Paspalum Turf Ecotypes

C.A. Cardona and R.R. Duncan

INTRODUCTION

Grasses in the genus *Paspalum* characteristically have evolved in and are adapted to extremely harsh environments. With over 400 species, *Paspalum* includes morphologically diverse plants ranging from bunch to prostrate-growing types and with variable leaf textures ranging from very coarse ornamental types [similar to St. Augustinegrass: *Stenotaphrum secundatum* (Walter) Kuntze] to fine-textured types resembling hybrid bermudagrass [*Cynodon dactylon* (L.) Pers. X *C. transvaalensis* Burtt-Davy] (Skerman and Riveros, 1989).

Paspalum vaginatum Swartz, or seashore paspalum (also called saltwater couch, siltgrass, sand knotgrass) is a rhizomatous-stoloniferous prostrate-growing turfgrass that evolved in wet, salinity-affected ecological zones (Morton, 1973). The grass is found predominately between 35° N-S latitudes in coastal domains. Although the center of origin for this grass has not been determined, South Africa and Argentina/Brazil are likely candidates.

Close relatives include dallisgrass (*Paspalum dilatatum* Poir.), which is a persistent bunch-type weedy species that plagues tropical and subtropical turf environments, and bahiagrass (*Paspalum notatum* Flugge), which is a bunch-type forage grass and stoloniferous turfgrass in southern tropical zones. A companion species is *Paspalum distichum* L. (or knotgrass, freshwater couch, eternity grass), which is a stoloniferous-rhizomatous grass that is found near fresh water ecozones and further inland than seashore paspalum.

Paspalum is a genetically complex genus with chromosome base number x=6 and 10 (Watson and Dallwitz, 1992). Diploids, tetraploids, and hexaploids are found with 2n=2x=20,40, 48, 50, 60, 63, and 80. Some species are apomictic (Chapman, 1992). Seashore paspalum is a sexually propagated diploid with predominantly 20 chromosomes (Burson, 1981); however, it is self-incompatible (Williams et al., 1994) and viable seed production ranges from 0–5% (R.R. Duncan, personal communication, 1995). This trait is found in other *Paspalum* species (Bennett and Bashaw, 1966) and this poor viable seed production characteristic in seashore paspalum is an obstacle to genetic diversification and improvement using conventional breeding methods.

Recovery of useful variability can be generated without sexual recombination by *in vitro* somaclonal variation (Larkin, 1987) and selection of somaclonal variants (Chaleff, 1983; Davies and Cohen, 1992; Duncan et al., 1995; Evans, 1989; Karp, 1995). Approximately 50 useful somaclonal variants have been released among various horticultural and agronomic crops (Duncan, 1996). Tissue culture traditionally consists of four steps: explant establishment of aseptic culture, initiation of callus, induction of embryoids, and regeneration of plants (Ahloowalia, 1984; Street, 1977). In *Paspalum*, immature inflorescences have been used as explants for induction of callus in 15 species (Bovo and Mroginsky, 1985) and in *P. dilatatum* Poir (Akashi and Adachi, 1992). In bahiagrass, embryogenic callus has been obtained from mature caryopses (Marousky and West, 1990), mature and immature embryos (Bovo and Mroginsky, 1989; Akashi et al., 1993), and leaf blades (Shatter et al., 1994). In *Paspalum almum* Chase, mature ovaries were the explant source (Bovo and Mroginsky, 1993). The objective of this study was to evaluate tissue culture–regenerated paspalum plants under turf conditions in the field and to determine if somaclonal variation of any turf traits can be used as a selection strategy.

MATERIALS AND METHODS

In Vitro Protocol

Immature inflorescences of nine seashore paspalum (hereafter referred to as paspalum) ecotypes (PI 509021—Argentina; HI-1, Mauna Key, K-3, K-7—Hawaii; Adalayd—Australia; PI 299042—Zimbabwe; S1PV-1—Sea Island, Georgia; AP-6—Ft. Myers, Florida) were used as the explant source for initiation of callus. The explants were surface-sterilized by soaking the entire inflorescence in 70% alcohol for 2 minutes, followed by soaking in a 10% chlorox solution for 2 minutes. All leaves were subsequently removed from the inflorescences and each spikelet in the raceme was sterilized in a 10% chlorox solution using agitation in a laminar flow hood. The spikelets were then rinsed four times with distilled water. After the last rinse, 30 explants per ecotype were cut into 10–15 mm long transverse sections for subsequent culture. The best medium for callus induction included 1 mg/L 2,4-D plus 5% coconut milk (Cardona, 1996). The basic medium consisted of Murashige and Skoog basal salts, 3% sucrose, and 1 mg/L 1000X Gamborgs B5 vitamins. The medium was sterilized by autoclaving at 121°C for 20 minutes. The cultures were maintained in the dark at 26–28°C for callus initiation and proliferation. Clumps of callus (2 to 5 mm) from each ecotype were replated to half-strength basic medium. The best medium for initiation of embryogenesis included 1 mg/L BAP plus 0.5–2 mg/L NAA. The cultures were subjected to a 16 hr light (46–60 µmol/sec) and 8 hr dark photoperiod. After 4 weeks, ecotype cultures producing shoots were cultured in half-strength, hormone-free medium with 8% sucrose for root proliferation.

Field Evaluation

Regenerated plantlets were transplanted to the field during 1993, 1994, and 1995 for turf trait evaluation. Regenerated plants were grown on a Cecil sandy clay loam (clayey, kaolinitic, thermic) at pH 6.0. Trait evaluations were made during 1994, 1995, and 1996.

Transformation Protocol

Highly embryogenic callus of PI 299042, HI-1, Mauna Key, and SIPV-1 were inoculated separately with two different binary vectors of *Agrobacterium tumefasciens*. Two cultures of *At* strain EHA105 were cotransformed with these two vectors:

(a) pBI 121 with NPTII (nos promoter) and GUS (CaMV 35s promoter)
(b) pGE 203 with NPTII (nos promoter), GUS (CaMV 35s promoter), and HPH (CaMV 35s promoter)

The inoculated callus was initially cultured for two days at 26°C in a callus induction medium, followed by sequential culturing in a shoot induction medium with 150 mg/L Claforan and transfer four days later to a medium of 100 mg/L each of Claforan and Kanamacin. Non-phenolic forming, active calli were reselected and transferred to fresh media plus selection agents on 30-day cycles.

RESULTS

Callus induction efficiency [percent of explants that produced viable callus [(total no. explants ÷ total – contaminated explants) × 100] ranged from 50% for PI 229042 to 39 (SIPV-1), 31 (PI 509021), 25 (K-7) to 12 (Mauna Key), 9 (AP-6), 7 (Adalayd), and 4% for HI-1. Ecotypes that were prolific producers of regenerated plants included HI-1, Mauna Key, PI 299042, K-3. K-7, SIPV-1, and AP-6 were intermediate, while PI 509021 and Adalayd were poor producers (Cardona, 1996).

SOMACLONAL VARIATION

Variability for genetic color, spread (diameter by growth rate over time), density (internode length), and winter hardiness (including early spring green-up) has been observed in Georgia turf evaluation plots at two field locations. A comparison of internode length means and ranges for parents and regenerants is found in Table 19.1. The means of the regenerants were greater than the parents for the five paspalum ecotypes; however, the regenerants had a wider range than the parents, reflecting both shorter and longer internode lengths. Selection of shorter internode somaclones was possible since about 7% (136 plants out of 1848) of the regenerated plants exhibited lengths less than their parents. As an example of somaclonal variation within one ecotype, ratings on spread for tissue culture–regenerated HI-1 ranged from 2 to 8 (8 = rapid growth and spread in diameter); for density, 2 to 8 (8 = most dense); and color, 5 to 8 (8 = darkest green).

Assessment of winter survival was made during July 1996 after low winter temperatures of –11.1 (1993), –14.4 (1994), –10.6 (1995), and –13.9°C (1996) were recorded in the field. A total of 2851 tissue culture–regenerated (TCR) plants from six paspalum ecotypes were evaluated. Finer-textured ecotypes HI-1 and Mauna Key had 28 and 24%, respectively, of the TCR plants killed by cold temperature, but 63 and 67% survived as healthy plants. Intermediate leaf texture types PI509021, K-3, and Adalayd had 53, 41, and 44%, respectively, of the TCR plants killed and 34, 52, and 32% healthy survivors. The coarse-textured type PI299042 lost 95% of the TCR plants to cold winter temperature and only two of the 43 survivors were in excellent condition. One group of TCR plants from PI299042 had no coarse-textured survivors, but had mutated to 15 (out of 61 total TCR plants) intermediate (Adalayd-type) texture types with improved cold tolerance.

Among the winter hardy survivors, approximately ten selections have been made, half of which broke winter dormancy prior to bermudagrass or zoysiagrass (*Zoysia* spp.) during early April 1996 when soil temperatures at the 10 cm depth were 10°C. The selections are being increased for large-scale turf evaluations during 1997.

Table 19.1. Somaclonal Variation for Internode Length (mm) Among 1848 Tissue Culture Regenerants.[a]

Paspalum	Parent			Regenerants		
Ecotype	Mean	Range	N	Mean	Range	N[b]
HI-1	7.4	6–10	20	9.5	2–26	565 (43)
PI 509021	7.6	5–11	20	9.0	3–20	421 (22)
K-3	9.3	7–12	20	10.6	4–30	400 (34)
Mauna Key	6.9	5–9	20	11.1	3–30	280 (12)
Adalayd	11.2	9–16	20	12.6	4–23	182 (25)

[a] Field data collected during 1995 in Georgia from 2-year-old clones.
[b] Number of regenerated plants with internode lengths less than the shortest internode length of the parents. 136 total plants.

Transformation

After 90 days on selection media, all calli inoculated with *At* strain pBI 121 were discarded with no results. Calli transformed with strain pGE203 showed active growth in the selection media (Table 19.3). One green plant from HI-1 grew extremely fast after emergence and was transferred to fresh media with 50 mg/L each of Claforan and Kanamacin to stimulate rooting. As soon as sufficient roots develop, gene expression evaluations will be conducted. A preliminary GUS expression test was performed, but results were negative. Polymerase chain reaction (PCR) tests will be performed using partial fragments of the transformation constructs as primers to identify whether the resistant plant is the product of transformation, an epigenic mutation, or a somaclonal variant resulting from the *in vitro* process.

CONCLUSIONS

The selection of tissue culture–regenerated somaclones for favorable turf traits has been possible due to induced or introduced variation during the *in vitro* process (Duncan, 1996). Over 2800 regenerated plants have been planted for turf evaluation in the field and over 100 selections exhibiting improved turf traits (short internode length, variable growth rates, improved winter hardiness) that are superior to the parents have been made. The selection for improved winter hardiness was especially encouraging, since the cold thermal threshold for paspalum has been determined to be –7 to –8°C (Campbell, 1979; Ibitayo et al., 1981; Fry, 1991). Similar results have occurred in wheat (*Triticum aestivum* L.) (Kendall et al., 1990) and melon (*Cucumis melo* L.) (Ezura et al., 1995). Additional wide-scale turf evaluations of the selected regenerants will be needed over several years to determine whether these improved ecotypes will eventually be released. This breeding strategy offers another method for

Table 19.2. Winter Survival Field Assessment of Paspalum Tissue Culture–Regenerated Plants in Georgia.[a]

Paspalum	Tissue Culture–Regenerated Plants			
Ecotype	Total	Dead	Poor Condition[b]	Survivors
HI-1	606	171 (28)[c]	52	383 (63)
PI 509021	409	216 (53)	52	141 (34)
K-3	432	177 (41)	29	226 (52)
Mauna Key	296	70 (24)	28	198 (67)
Adalayd	207	92 (44)	48	67 (32)
PI 299042	901	858 (95)	—	43 (5)
Total	**2851**	**1584 (56)**	**209 (7)**	**1058 (37)**

[a] At Griffin, Georgia; coldest winter temperatures were –11.1°C (1993), –14.4°C (1994), –10.6°C (1995), and –13.9°C (1996). Data recorded July, 1996. Cold thermal threshold range of –7 to –8°C has been reported for paspalum (Campbell, 1979; Ibitayo et al., 1981; Fry, 1991).

[b] Plants were small and struggling to recover from the winter cold temperatures.

[c] Number in parentheses indicates percentage of total number of plants.

Table 19.3. Growth of Paspalum Callus Inoculated with *Agrobacterium tumefasciens* Strain p GE 203.

	Paspalum Ecotype			
Results	PI 299042	HI-1	Mauna Key	SIPV-1
Number of active growth callus (foci)	6	19	8	3
Developmental status	Formed roots only	Produced 4 albino and 1 green plant	Formed roots only	Formed roots only

creating variability in a sexually incompatible grass involving multiple turf traits and selection in the field. The potential for this environmentally friendly turfgrass (Duncan, 1996a,b) remains promising for the future.

REFERENCES

Ahloowalia, B.S. 1984. Plant cell culture in forage grasses. In *Handbook of Plant Cell Culture*, Vol. 3. Ammirato, P.V., D.A. Evans, N.R. Sharp, and P. Yamada (Eds.). McMillan Publishers Co., NY, pp. 91–125.

Akashi, R. and T. Adachi. 1992. Somatic embryogenesis and plant regeneration from cultured inflorescences of *Paspalum dilatatum* Poir. *Plant Science*, 82:213–218.

Akashi, R., A. Hashimoto, and T. Adachi. 1993. Plant regeneration from seed derived embryogenic callus and cell suspension cultures of bahiagrass (*Paspalum notatum*). *Plant Science*, 90:73–80.

Bennett, H.W. and E.C. Bashaw. 1966. Interspecific hybridization in *Paspalum* spp. *Crop Science*, 6:52–54.

Bovo, O.A. and L.A. Mroginski. 1985. Tissue culture in *Paspalum (Gramineae)*. Plant regeneration from cultured inflorescences. *Journal of Plant Physiology*, 124:481–492.

Bovo, O.A. and L.A. Mroginski. 1989. Somatic embryogenesis and plant regeneration from cultured mature and immature embryos of *Paspalum notatum (Gramineae). Plant Science,* 85:217–223.

Bovo, O.A. and L.A. Mroginski. 1993. Obtencion de plantas de *Paspalum almum (Gramineae)* a partir del cultivo *in-vitro* de ovarios jovenes. *Phyton*, 43:29–34.

Burson, B.L. 1981. Genome relationship among four diploid *Paspalum* species. *Botanical Gazette*, 142:592–596.

Campbell, W.F. 1979. Futurf: a new grass for Utah's Dixie. *Utah Science,* June, 32–33.

Cardona, C.A. 1996. Development of a tissue culture protocol and low temperature tolerance assessment in *Paspalum vaginatum* Sw. University of Georgia, Ph.D. Dissertation, 91 pp.

Chaleff, R.S. 1983. Isolation of agronomically useful mutants from plant cell cultures, *Science*, 219:676–682.

Chapman, G.P. 1992. Apomixis and evolution, in *Grass Evolution and Domestication*, Chapman, G.P. (Ed.). Cambridge University Press, United Kingdom. pp. 138–155.

Davies, L.J. and D. Cohen. 1992. Phenotypic variation in somaclones of *Paspalum dilatatum* and their seedling offspring, *Canadian Journal of Plant Science*, 72:773–784.

Duncan, R.R., R.M. Waskom, and N.W. Nabors. 1995. *In vitro* screening and field evaluation of tissue-culture-regenerated sorghum [*Sorghum biolor* (L.) Moench] for soil stress tolerance, *Euphytica*, 85:373–380.

Duncan, R.R. 1996a. Seashore paspalum: the next-generation turf for golfcourses. *Golf Course Management.* April, pp. 49–51.

Duncan, R.R. 1996b. The environmentally sound turfgrass of the future—seashore paspalum can withstand the test. *U.S. Golf Association Green Section Record,* 34(1):9–11.

Duncan, R.R. 1996. Tissue culture induced variation and crop improvement. *Advances in Agronomy*, 58:201–240.

Evans, D.A. 1989. Somaclonal variation: genetic basis and breeding applications. *Genetics*, 5:46–50.

Ezura, H., H. Amagai, I. Kikuta, M. Kubota, and K. Oosawa. 1995. Selection of somaclonal variants with low-temperature germinability in melon (*Cucumis melo* L.). *Plant Cell Reports*, 14:684–688.

Fry, J. 1991. Freezing resistance of southern turfgrasses. *Lawn and Landscape Management*, January, pp. 48–51.

Ibitayo, O.O., J.D. Butler, and M.J. Burke. 1981. Cold hardiness of bermudagrass and *Paspalum vaginatum* Sw. *Horticultural Science*, 16:683–684.

Karp, A. 1995. Somaclonal variation as a tool for crop improvement, *Euphytica*, 85:295–302.

Kendall, E.J., J.A. Qureshi, K.K. Kartha, N. Leung, N. Chevrier, K. Caswell, and T.H.H. Chen. 1990. Regeneration of freezing-tolerant spring wheat (*Triticum aestivum* L.) plants from cryoselected callus. *Plant Physiology*, 94:1756–1762.

Larkin, P.J. 1987. Somaclonal variation: history, method, and meaning. *Iowa State Journal of Research*, 61:393–434.

Marousky, F.J. and S.H. West. 1990. Somatic embryogenesis and plant regeneration from cultured mature caryopses of bahiagrass (*Paspalum notatum* Fluegge). *Plant Cell and Tissue Organ Culture*, 20:125–129.

Morton, J.F. 1973. Salt-tolerant siltgrass (*Paspalum vaginatum* Sw.) in *Proceedings of the Florida State Horticultural Society*, 6–8 November 1973, 86:482–490.

Shatters, R.G., Jr., R. Wheeler, and S.H. West. 1994. Somatic embryogenesis and plant regeneration from callus cultures of Tifton 9 bahiagrass. *Crop Science*, 34:1378–1384.

Skerman, P.J. and F. Riveros. 1989. *Tropical Grasses*. Food and Agricultural Organization of the United Nations, Rome, Italy, FAO Plant Production and Protection Series, No. 23, p. 119, pp. 565–568.

Street, H.E. 1977. Embryogenesis and chemically induced organogenesis. In *Plant Cell and Tissue Culture: Principles and Application*. Sharp, W.R., P.O. Larson, E.F. Paddock, and V. Ragharan (Eds.). Ohio State University Press, Columbus, OH. pp. 123–153.

Watson, L. and M.F. Dalliwitz. 1992. *The Grass Genera of the World*. C.A.B. International, Wallingford, Oxon, United Kingdom, pp. 672–674.

Williams, E.G., A.E. Clarke, and R.B. Knox. 1994. *Genetic Control of Self-Incompatibility and Reproductive Development in Flowering Plants*. Kluwer Academic Publishers, Dordrecht, Holland.

Part 5

Future Perspectives

Chapter 20

Molecular Biology: A Whole-Plant Physiologist's Journey to the Dark Side

D.C. Bowman

First, let me admit that I am not a molecular biologist. I have never taken a course in the subject, and what little I know I've picked up from colleagues infinitely more capable than I. But I do recognize the power in the techniques. I do see the potential to address problems in turf physiology and management that we otherwise could not. And I see opportunities for technology transfer and product development that may return enough in royalties or licensing fees to keep some of us working for a while. Following are two examples of how molecular biology is being incorporated into traditional turfgrass research.

Nitrogen is a primary determinant of agricultural crop productivity and ornamental crop quality. It is also a relatively expensive input for many crops. Unfortunately, N uptake efficiency by crops is often quite low, rarely exceeding 60% absorption of the N applied. This is due to several physical and biological processes, such as leaching, gaseous losses, and microbiological immobilization, which compete with the crop for N. A detailed understanding of both root absorption and the competing processes should ultimately result in more efficient N absorption, more profitable plant production, and less NO_3 leaching into and contaminating surface and groundwaters.

Much of my research has characterized nitrogen uptake by both cool- and warm-season turfgrasses at the whole plant level. Several years ago I decided to try to identify the putative carrier protein involved in nitrate transport into roots to better understand the uptake process, a decision that ultimately led me to the dark side. *Note:* For purposes of selling this idea to funding agencies, it is always important to emphasize that identifying the carrier protein is critical if we hope to improve nitrate absorption, and thus reduce nitrate leaching.

Nitrate uptake is extremely well suited for study because the process is substrate inducible, a characteristic that is fundamental to our strategies for identifying the transporter. In seedling roots never exposed to nitrate, or mature roots that have been deprived of nitrate for a period as short as 24 hours, uptake rates are very low (Figure 20.1). Exposing the roots to nitrate presumably causes transcription and translation of gene(s) that code for a nitrate transport protein. This protein is then targeted to the plasmalemma, where it acts to facilitate nitrate movement across the membrane. Uptake rates increase fairly quickly after exposure of roots to nitrate, and reach a maximum after 8 to 12 hours.

As a plant physiologist with just enough background in biochemistry to get in trouble, I initially attempted to visualize membrane proteins unique to induced root tissue using PAGE electrophoresis of fractions enriched for plasmalemma. Although unique bands were often seen on the gels, these were not reproducible. Since membrane preparation by aqueous two-phase partitioning is somewhat time-consuming, it is possible that the carrier protein had degraded prior to electrophoresis. Two-dimensional gels were no more successful at resolving this protein.

A second biochemical approach was then attempted. Phenylglyoxal is an inhibitor of certain proteins involved in anion binding or transport. This compound reportedly binds irreversibly to arginine residues, which are often functionally involved in anion transport. We hypothesized that the nitrate binding site contained essential arginines, support for which was found in research documenting that phenylglyoxal dramatically inhibited nitrate uptake. Our idea was to use phenylglyoxal to tag the carrier protein. Unfortunately, the inhibitor would also bind to any other membrane protein having an exposed arginine residue. Several reports indicated that the essential arginine residues of some proteins could be protected from phenylglyoxal inhibition by their anion substrates. We reasoned that if we could protect the nitrate carrier specifically by including nitrate in the reaction, we could label the non-target proteins with "cold" phenylglyoxal. We could then come back with a minus-nitrate reaction containing ^{14}C-labeled inhibitor, and preferentially label the now unprotected nitrate carrier. Labeled proteins could then be separated on a gel and visualized. For this strategy to work, phenylglyoxal binding would have to be irreversible, as reported in the literature for other systems, and the binding site of the carrier would have to be protected by nitrate. In preliminary characterization of the inhibitor, however, we found evidence that the binding was at least partly reversible (Figure 20.2), and that nitrate did not protect the carrier from inhibition (Table 20.1). After much preliminary work, we hit a dead end. And we realized that some questions might not readily surrender to traditional biochemical approaches. Having little background in molecular biology, I was

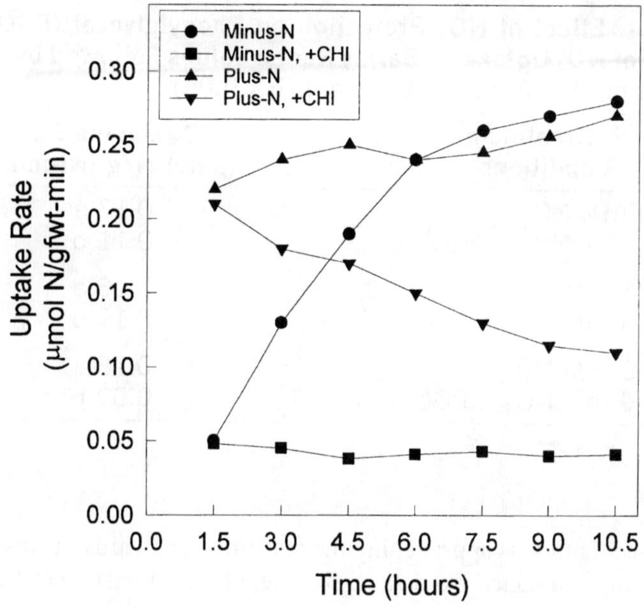

Figure 20.1. Induction of NO_3 uptake in N-deficient barley roots compared to fully induced uptake in plus-N roots. Cyclohexamide (CHI) was supplied during the entire period at 5 μg/mL.

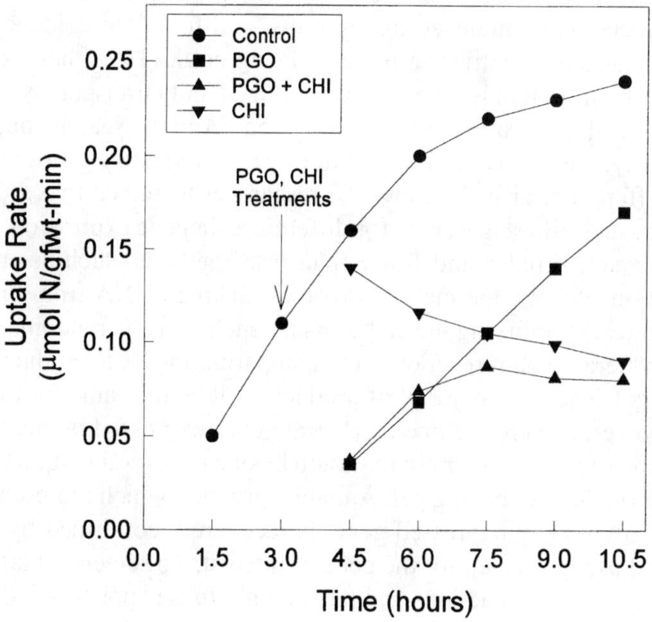

Figure 20.2. Effect of phenylglyoxal (PGO, 4 m*M*) and cyclohexamide (CHI, 5 μg/mL) on NO_3 uptake by N-deficient barley roots.

Table 20.1. Effect of NO_3 Protection on Phenylglyoxal (PGO, 2 m*M*) Inhibition of NO_3 Uptake by Barley Roots. Values Followed by the Same Letter are not Significantly Different ($P < 0.05$).

Pretreatment Conditions	NO_3 Uptake Rate (µmol N/g fwt-min)
1 m*M* NO_3	0.17 a
1 m*M* NO_3 + PGO	0.11 b
25 m*M* NO_3	0.16 a
25 m*M* NO_3 + PGO	0.10 b
50 m*M* NO_3	0.16 a
50 m*M* NO_3 + PGO	0.09 b

unable to see that I was approaching the question backwards. It was only after much help and guidance from several molecular biologists that I concluded the most direct assault on the protein actually targeted the gene.

In 1993, Tsay et al.'s group reported their discovery of a gene from *Arabidopsis* encoding a nitrate transporter. The obvious question was whether a homologous transport gene existed in monocots other than maize. We generated a probe from their published sequence using PCR and screened an *Arabidopsis* cDNA library for the clone. We probed a Northern blot from nitrate-induced and uninduced barley roots, and found no evidence that the gene was induced by nitrate in barley. Huang et al. (1996) have since concluded that their protein is involved in the low-affinity transport system rather than the inducible, high-affinity uptake system. And the search continues.

Presently, we are using differential display to identify genes induced by nitrate. Differential display allows the researcher to screen two different organisms, tissues, life stages, etc., for differences in gene expression. This procedure is much simpler and faster than past methods, such as subtractive hybridization. Briefly, the method involves making cDNA from the two tissues of interest, amplifying the cDNA using anchored and random primers by PCR (polymerase chain reaction), and comparing the PCR products on a sequencing gel. The vast majority of products will be the same for the two tissues. However, if there are one or several genes expressed in one tissue that aren't expressed in the other, there should be one to several unique PCR products found on the sequencing gel. A unique product, which represents a small section of a uniquely expressed gene, is recovered, confirmed by RT-PCR, cloned, and used to screen for the gene of interest. To screen out artifacts, we typically run parallel reactions and pursue only those "positives" that appear in both gels. We are using differential display to successfully screen for genes involved in nickel uptake and tolerance, and are in the process of applying it to the nitrate transporter in turfgrasses.

A second project in which molecular techniques are being used falls under the category of turf management. Specifically, we are interested in evaluating strategies to convert an existing Penncross putting green to one of the newer varieties. Many superintendents have decided to switch to an improved variety, but are reluctant to close their greens during renovation. The challenge then is to develop a program to facilitate a gradual conversion from one bentgrass to another. Several management practices, including verticutting, shallow core cultivation, and use of growth regulators, are obvious choices to consider to improve overseeding success. But how can success, or failure, of our treatments be measured? Isozyme profiles are too variable and often lack adequate resolution. We have decided to use RAPD markers to determine the population makeup in our overseeding treatments. Finding suitable markers has been relatively easy, since we only need to distinguish between Penncross and one other variety, rather than between all the numerous varieties on the market. And the technique itself has become routine enough so that an undergraduate can handle the analyses.

It is apparent from the interest in this meeting that molecular biology is having an impact on turfgrass science. Many turf researchers have embraced the new technology and are making advances. But we are faced with several challenges. How can we educate our students and industry to foster their understanding and support? How can we best work with private industry, since in many cases they hold most of the cards? And how can we generate enough financial support to develop successful programs? These questions must be answered for turfgrass biotechnology to fulfill its promise.

REFERENCES

Huang, N-C., C-S. Chiang, N.M. Crawford, and Y-F. Tsay. 1996. CHL1 encodes a component of the low-affinity nitrate uptake system in *Arabidopsis* and shows cell type-specific expression in roots. *Plant Cell* 8:2183–2191.

Tsay, Y-F., J.I. Schroeder, K.A. Feldmann, and N.M. Crawford. 1993. A herbicide sensitivity gene *CHL1* of *Arabidopsis* encodes a nitrate-inducible nitrate transporter. *Cell* 72:705–713.

Index

A

B

C

D

E

F

I

J

K

L

P

Q

R

S

W

X

Y

Z